André Kappes

Monodromy Representations and Lyapunov Exponents of Origamis

Monodromy Representations and Lyapunov Exponents of Origamis

by

André Kappes

KIT Scientific Publishing

Dissertation, Karlsruher Institut für Technologie
Fakultät für Mathematik
Tag der mündlichen Prüfung: 25. Mai 2011
Referenten: JProf. Dr. Gabriela Weitze-Schmithüsen
Prof. Dr. Frank Herrlich
Prof. Dr. Martin Möller, Goethe-Universität Frankfurt

Impressum

Karlsruher Institut für Technologie (KIT)
KIT Scientific Publishing
Straße am Forum 2
D-76131 Karlsruhe
www.ksp.kit.edu

KIT – Universität des Landes Baden-Württemberg und nationales
Forschungszentrum in der Helmholtz-Gemeinschaft

KIT Scientific Publishing 2011
Print on Demand

ISBN 978-3-86644-751-6

Preface

An *origami* is a compact Riemann surface X, which is tiled by finitely many Euclidean unit squares. An example is given in Figure 0.1. Away from the vertices, the tiling provides a particular atlas for X: Locally, its transition maps are translations $z \mapsto z + c$, $c \in \mathbb{C}$. More generally, any finite collection of polygons in the Euclidean plane, whose sides can be paired by translations, gives rise to a compact Riemann surface with such a *translation structure* ω. The pair (X, ω) is called a *translation surface.*

If we apply the linear action of $A \in \mathrm{SL}_2(\mathbb{R})$ to the collection of polygons, we still can pair the sides accordingly, but the translation structure and even the complex structure of the deformed Riemann surface will usually differ from the original one. However, it may happen that there is a way to cut and reglue the polygons to obtain the original collection. In this case, we can find a homeomorphism $f : X \to X$, which is affine w. r. t. to the translation structure and whose matrix part is A. The affine homeomorphisms of X assemble to a group $\mathrm{Aff}(X, \omega)$; the *Veech group* $\Gamma(X, \omega)$ is the discrete subgroup of $\mathrm{SL}_2(\mathbb{R})$ of matrix parts of affine homeomorphism.

The generic translation surface will admit (almost) no affine homeomorphisms. A translation surface that has many is called *Veech surface.* The most basic examples of Veech surfaces are origamis: Their Veech groups are always commensurable with $\mathrm{SL}_2(\mathbb{Z})$.

If a translation surface (X, ω) of genus g has a big Veech group, the isomorphism classes of Riemann surfaces obtained as affine deformations of (X, ω) trace out an algebraic curve in the moduli space $\mathcal{M}_g$ of compact Riemann surfaces of genus g. Such curves are called Teichmüller curves, since they are totally geodesic for the Teichmüller metric on $\mathcal{M}_g$. A Teichmüller curve coming from a Veech surface (X, ω) is uniformized by $\mathbb{H}/\Gamma(X, \omega)$, and the Riemann surfaces parametrized by its points can be assembled to

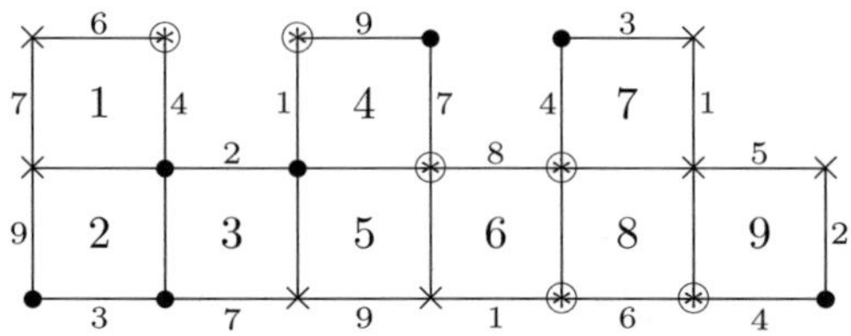

Figure 0.1: *An origami. The number at a side indicates which square is adjacent.*

a family of curves $\phi : \mathcal{X} \to \mathcal{C} = \mathbb{H}/\Gamma$ after passing to a suitable finite-index subgroup Γ of $\Gamma(X,\omega)$.

In this thesis, we study the action

$$\rho : \mathrm{Aff}(X,\omega) \to \mathrm{Aut}(H^1(X,\mathbb{Z})), \quad f \mapsto (f^{-1})^*$$

for Veech surfaces (X,ω), with particular emphasis on the ones coming from origamis. The restriction of ρ to $\Gamma = \pi_1(\mathcal{C})$ is the *monodromy action* of the family ϕ. This group action is equally described by the *local system* $R^1\phi_*(\mathbb{Z})$ on $\mathcal{C}$. As an additional datum, this local system carries a *polarized variation of Hodge structures* (pVHS) $\phi_*\omega_{\mathcal{X}/\mathcal{C}} \subset R^1\phi_*(\mathbb{Z}) \otimes_{\mathbb{Z}} \mathcal{O}_{\mathcal{C}}$. It was proved by M. Möller [Möl06] that the pVHS on a Teichmüller curve is characterized by a particular splitting into sub-pVHS. In the case of an origami, the pVHS splits over $\mathbb{Q}$ into $\mathbb{L} \oplus \mathbb{M}$, where $\mathbb{L}$ is a sub-VHS of rank 2, corresponding to the Fuchsian representation $\Gamma \to \mathrm{SL}_2(\mathbb{R})$. Our interest lies in the description of the remaining part $\mathbb{M}$.

We exhibit two basic concepts leading to sub-pVHS of $\mathbb{M}$:

Theorem 9.3 *If $p : X \to Y$ is a covering map to another Veech surface Y, compatible with the translation structures, then up to adjusting Γ, the pVHS $R^1\phi_*(\mathbb{Z})$ has a sub-pVHS of rank $2g(Y)$ induced by p.*

Theorem 9.6 *Let* $\mathrm{Aut}(X,\omega)$ *be the subgroup of biholomorphic affine homeomorphisms. Up to passing to a finite-index subgroup of Γ, ev-*

ery $\mathrm{Aut}(X,\omega)$*-isotypic component of* $H^1(X,\mathbb{R})$ *furnishes a sub-pVHS of* $R^1\phi_*(\mathbb{R})$.

We apply the second concept to the origami $\widetilde{\mathbf{St}}_3$ of degree 108 found by F. Herrlich [Her06]. Using a formula of C. Chevalley and A. Weil, we determine the decomposition of $H^1(\widetilde{\mathbf{St}}_3,\mathbb{C})$ into isotypic components in Proposition 9.11.

As to the first concept, we compute the splitting of the pVHS for a number of origamis in Section 9.4. We first carry out the computations for the toy model origami $\mathbf{L}_{2,2}$, and then proceed to the discussion of the origami $\mathbf{M}$ shown in Figure 0.1, which is a cover of two origamis in the $\mathrm{SL}_2(\mathbb{Z})$-orbit of $\mathbf{L}_{2,2}$. We show that the pVHS of $\mathbf{M}$ decomposes completely into irreducible rank 2-summands. For a 3-fold cover $\mathbf{N}_3$ of $\mathbf{M}$, we obtain a decomposition into irreducible rank 2-summands and some unitary summands.

Our examples stem from the first member of an infinite family $\mathbf{N}_n$ of origamis described in Section 3.2. This family is special in that its members admit no non-trivial translations, i. e. affine biholomorphisms with derivative I, but have the maximum possible number of affine homeomorphisms, i. e. their Veech group is $\mathrm{SL}_2(\mathbb{Z})$. Theorem 3.3 summarizes the properties of the origamis $\mathbf{N}_n$.

The monodromy action also figures in an interesting dynamical system, the *Kontsevich-Zorich cocycle* [Zor96], which is a cocycle over the Teichmüller flow on $\mathcal{C}$. The *Lyapunov exponents* associated with this cocycle describe the mean growth rate of cohomology vectors along a generic geodesic in $\mathcal{C}$. The Lyapunov spectrum, i. e. the collection of the Lyapunov exponents, is given as

$$1 = \lambda_1 > \lambda_2 \geq \cdots \geq \lambda_g \geq 0 \geq -\lambda_g \geq \cdots \geq -\lambda_2 > -\lambda_1 = -1.$$

It is in general not known, which numbers $\lambda_2,\dots,\lambda_g$ can occur. However the sum

$$1 + \lambda_2 + \cdots + \lambda_g$$

is given by an algebraic quantity: As shown by several authors [Kon97], [For02], it is the quotient of degrees of certain line bundles. A variation on this theme is obtained when we are able to find a sub-pVHS $\mathcal{L}^{1,0} \subset \mathcal{L} = \mathbb{L} \otimes_{\mathbb{C}} \mathcal{O}_{\mathcal{C}}$ of rank 2 [BM10b]. In this case, one finds a single non-negative Lyapunov exponent of the spectrum

$$\lambda_{\mathbb{L}} = \frac{\deg(\mathcal{L}^{1,0})_{\text{ext}}}{2g(\mathcal{C}) - 2 + |\overline{\mathcal{C}} \setminus \mathcal{C}|}. \tag{0.1}$$

Following an idea of M. Möller, we show in Section 9.3 that it is possible to compute the numerator of the fraction in (0.1) with the help of the period map if one knows explicitly the action of Γ on the associated 2-dimensional subspace of the first cohomology of X.

Theorem 9.18 *Suppose $U \subset H^1(X, \mathbb{Z})$ is a Γ-invariant subspace of rank 2 whose associated local system carries a sub-pVHS of $R^1\phi_*(\mathbb{Z})$. If Γ acts on U by a finite-index subgroup $\rho(\Gamma)$ of $\mathrm{SL}_2(\mathbb{Z})$, then the associated non-negative Lyapunov exponent is given by*

$$\lambda_U = \frac{\deg(p)\,\mathrm{vol}(\mathbb{H}/\rho(\Gamma))}{\mathrm{vol}(\mathbb{H}/\Gamma)},$$

where $p : \mathbb{H}/\Gamma \to \mathbb{H}/\rho(\Gamma)$ is the period mapping associated with the sub-pVHS.

We carry out the computation of the Lyapunov exponents for our examples and are able to completely determine their Lyapunov spectrum (see Corollary 9.26 and Corollary 9.28).

Finally, we find a sub-pVHS of the pVHS of the origami $\mathbf{M}$, for which we show in Proposition 9.34 that it does not occur in genus 2. To this end, we introduce the notion of *period data* in Section 9.5.

Structure of the thesis

This thesis is structured as follows:

In Chapter 2 we give an overview of the basic concepts. We discuss translation surfaces, Teichmüller curves, moduli spaces, and origamis. This theory is widely known, and we will only give some proofs, either for clarification or because we do not know a reference.

In Chapter 3 we first present Herrlich's construction of characteristic origamis and then deduce from his result the existence of an infinite family of origamis with many affine symmetries but no non-trivial translations. We close this section by pointing out a funny construction by M. Schmoll, leading also to origamis with Veech group $\mathrm{SL}_2(\mathbb{Z})$.

Chapter 4 subsumes basic facts about the cohomology of a compact Riemann surface (or more generally, a Kähler manifold).

In Chapter 5 we discuss the notion of a family of curves. In particular, we include a condition on how to choose the finite unramified cover $\mathbb{H}/\Gamma \to \mathbb{H}/\Gamma(X,\omega)$ of a Teichmüller curve in order that there exist a family of curves over $\mathbb{H}/\Gamma$.

Chapter 6 recounts the threefold description of monodromy actions, local systems and vector bundles with a flat connection given by P. Deligne, enriched by some specializations to the case of Teichmüller curves.

In Chapter 7 we present the abstract notion of a Hodge structure and of a variation of Hodge structures. We construct the period mapping and explain how Hodge structures and abelian varieties are related. Moreover, we recall M. Möller's characterization of the pVHS on a Teichmüller curve.

Chapter 8 is a summary of the aspects in dynamical systems that we need: We discuss Oseledet's Theorem on the existence of Lyapunov exponents and then define the multiplicative cocycle that we are interested in, the Kontsevich-Zorich cocycle.

Chapter 9 contains the results stated above.

There are numerous people without whom this thesis would never have been finished. First and foremost, I would like to thank my advisor *JProf. Dr. Gabriela Weitze-Schmithüsen* for her permanent encouragement and her support in innumerous occasions, in particular during the last couple of stressful days before the deadline. She did everything to give me the most possible amount of time, while at the same time she was proofreading the manuscript very carfully. Many thanks go equally to the two Korreferenten *Prof. Dr. Frank Herrlich* and *Prof. Dr. Martin Möller*. The first one has a famous always open door (which is now competed by Gabi's); I am grateful to him for proofreading and many helpful discussions and his continuous support during the last years. The second one helped me find the way through variations of Hodge structures and period mappings in various very helpful email conversations. Special thanks go to *Prof. Alex Eskin* for giving me the opportunity of staying at the University of Chicago for six months.

Many thanks go also to *Myriam Finster*, who was always ready to help me out when I had some programming or mathematical issues and to *Florian Nisbach* who proofread parts of my thesis and who always lends me his ear when I have a question for him. *Stefan Kühnlein* was always willing to be bothered with my problems. I thank him especially for sharing his knowledge on representation theory. Moreover, my thanks go to all other past and present members of the famous "Kaffeerunde": to *Karsten Kremer*, whose thesis proves to be incredibly useful; to *Oliver Bauer*, whose thesis was a good starting point to explore monodromy representations of origamis, to *Ute Hoffmann* for her help with all organizational problems, and to all others for the very cordial and pleasant working atmosphere.

My stay in Chicago was made possible with financial support from the *Karlsruhe House of Young Scientists*; I thank them heartly for having given me this opportunity. Part of this research was done, while I was employed in the research project "Mit Origamis zu Teichmüllerkurven im Modulraum", led by JProf. Dr. Gabriela Weitze-Schmithüsen and commissioned by the *Landesstiftung Baden-Württemberg.*

Finally, and most importantly, I am indebted to my parents *Waltraud and Wolfgang Kappes* and to *Andrea Schumm* for their continuous support and love.

Karlsruhe, August 2011

Contents

1 Notations and Conventions **1**

2 Teichmüller Curves **3**
2.1 Teichmüller Spaces and Moduli Spaces of Curves 3
2.2 Translation Surfaces 6
2.3 Teichmüller Curves 8
2.4 Origamis 11

3 Characteristic Origamis **23**
3.1 Characteristic Origamis 23
3.2 An Infinite Series of Modular Origamis 24
3.3 Schmoll's Modular Fibers 29

4 Homology and Cohomology **31**
4.1 Singular Homology and Cohomology 31
4.2 De Rham Cohomology 33
4.3 Hodge Decomposition 34
4.4 Riemann Surfaces 36

5 Families of Curves **39**

6 Local Systems, Monodromy Representations, Flat Vector Bundles **43**
6.1 Local Systems 43
6.2 Monodromy Representation 45
6.3 Vector Bundles with a Flat Connection 47

7 Polarized Variations of Hodge Structures **51**
7.1 Hodge Structures 51
7.2 Variations of Hodge Structures 57
7.3 Variations of Hodge Structures and Families of Curves . . . 60

7.4 Characterization of Teichmüller Curves by their VHS 67

8 Multiplicative Ergodic Theory 71
8.1 Oseledets' Theorem 71
8.2 The Kontsevich-Zorich Cocycle 72
8.3 The Lyapunov Spectrum of Teichmüller Curves 77

9 Splitting the Hodge Bundle over a Teichmüller curve 81
9.1 Coverings 82
9.2 Representations 84
9.3 Rank 2 90
9.4 Examples 100
9.5 Equivalence of Rank 2-Period Mappings 120

A Appendix 127

Bibliography 129

Index 137

1 Notations and Conventions

Before we start, let us fix some notations that we use.

By a ring, we shall always understand a commutative ring with unity.

The Fundamental Group. Let X be a topological space and let $x \in X$. The fundamental group $\pi_1(X, x)$ of X based in x is the group of homotopy classes of continuous paths $\alpha : [0, 1] \to X$ starting and ending in x. To be precise, if we are given $a, b \in \pi_1(X, x)$, represented by α and β, then ab shall be the path obtained by *first* running along β and *then* running along α. We will disobey to this convention only in Chapter 9, where we use in some places the opposite group $\pi_1^{op}(X, x)$.

Riemann Surfaces and Algebraic Curves. A Riemann surface is a connected, 1-dimensional complex manifold. If considered as 2-dimensional real manifold, then it shall always be equipped with its natural orientation coming from the complex structure. Recall that there is an equivalence of categories between non-singular, projective algebraic curves over $\mathbb{C}$ and compact Riemann surfaces. Therefore, the term compact Riemann surface and algebraic curve (meaning non-singular, projective algebraic curve over $\mathbb{C}$) will be used synonymously.

Sheaves. If X is a topological space and if M is a set, then M_X shall denote the constant sheaf associated to M. We will often drop the subscript if the context is clear.

In the following let $\mathbb{K} = \mathbb{R}$ or $\mathbb{C}$. For a smooth manifold X, let $\mathrm{C}^\infty_{X,\mathbb{K}}$ denote the sheaf of smooth functions on X with values in $\mathbb{K}$. By a differentiable real (respectively complex) vector bundle over a smooth manifold X, we shall understand a locally free $\mathrm{C}^\infty_{X,\mathbb{R}}$-module (resp. $\mathrm{C}^\infty_{X,\mathbb{C}}$) $\mathcal{V}$ of finite type,

i. e. every point has a neighborhood U, where $\mathcal{V}_{|U} \cong (\mathrm{C}^\infty_{U,\mathbb{R}})^n$ (respectively $\mathcal{V}_{|U} \cong (\mathrm{C}^\infty_{U,\mathbb{C}})^n$) for some fixed n, called the rank of $\mathcal{V}$.

In the same way, if X is a complex space and if $\mathcal{O}_X$ is the sheaf of holomorphic functions on X, then we define a holomorphic vector bundle to be a locally free $\mathcal{O}_X$-module of finite type.

The fiber of a vector bundle $\mathcal{V}$ over X above the point $x \in X$ is the (real or complex) vector space

$$\mathcal{V}_x := \mathcal{V}_{(x)} \otimes_{\mathrm{C}^\infty_{(x)}} \mathrm{C}^\infty_{(x)} / m_{(x)},$$

where $\mathcal{V}_{(x)}$ is the stalk of $\mathcal{V}$ at x, $\mathrm{C}^\infty_{(x)}$ is the stalk of $\mathrm{C}^\infty_{X,\mathbb{K}}$ at x, and $m_{(x)} \subset \mathrm{C}^\infty_{(x)}$ is the maximal ideal of functions vanishing at x.

Replacing $\mathrm{C}^\infty_{X,\mathbb{K}}$ by $\mathcal{O}_X$ in the above definition, we obtain the definition of a fiber of a holomorphic vector bundle on a complex space.

Contrary to this notation, in general, the stalk of a sheaf $\mathcal{F}$ on a topological space X in the point $x \in X$ is denoted by $\mathcal{F}_x$.

2 Teichmüller Curves

2.1 Teichmüller Spaces and Moduli Spaces of Curves

We recall some generalities on the Teichmüller space and the moduli space of curves. They can be found e. g. in [Hub06], [Mas09], [EE69], [HM98], to list only some references.

Let S be a compact, smooth, oriented, connected 2-dimensional manifold of genus $g \geq 0$ with a set Σ of $n \geq 0$ marked points. We always assume that $3g - 3 + n > 0$. Let $\mathcal{J}(S)$ be the set of complex structures on S. $\mathcal{J}(S)$ can be described as the set of endomorphisms $J : TS \to TS$ of the tangent bundle TS with $J^2 = -\operatorname{id}$, such that the orientation induced by J coincides with the one given on S. $\mathcal{J}(S)$ itself is endowed with a topology and a complex structure (see [EE69]).

The group $\operatorname{Diffeo}^+(S, \Sigma)$ is the group of orientation-preserving diffeomorphisms $f : S \to S$, such that $f_{|\Sigma} = \operatorname{id}_\Sigma$. It acts on $\mathcal{J}(S)$; the normal subgroup $\operatorname{Diffeo}^0(S, \Sigma)$ of diffeomorphisms homotopic to the identity (through homotopies fixing Σ) acts freely. The quotient of $\mathcal{J}(S)$ by the action of $\operatorname{Diffeo}^0(S, \Sigma)$ is the *Teichmüller space* $\mathcal{T}(S, \Sigma)$. Let us fix a complex structure j_0, and let us also denote the corresponding compact Riemann surface (S, j_0) by S. Then $\mathcal{T}(S, \Sigma)$ can be identified with equivalence classes of pairs (X, m), where X is a compact Riemann surface of genus g, and $m : S \to X$ is an orientation-preserving diffeomorphism, called *marking*. Here, $(Y, n) \sim (X, m)$ if there exists $\phi \in \operatorname{Diffeo}^0(S, \Sigma)$ and $h : X \to Y$ a biholomorphic map such that

$$\begin{array}{ccc} S & \xrightarrow{m} & X \\ {\scriptstyle\phi}\downarrow & & \downarrow{\scriptstyle h} \\ S & \xrightarrow[n]{} & Y \end{array}$$

commutes. Note that Teichmüller space has a base point (S, id). If we do not care about the base point, we will simply write $\mathcal{T}_{g,n}$ in place of $\mathcal{T}(S, \Sigma)$. Also, $\mathcal{T}(S) = \mathcal{T}(S, \emptyset)$ and $\mathcal{T}_g = \mathcal{T}_{g,0}$.

An alternative and more involved way to describe the Teichmüller space $\mathcal{T}(S, \Sigma)$ is the use of quasiconformal homeomorphisms as markings. The definition is parallel to the one given above. It allows to represent $\mathcal{T}(S, \Sigma)$ as the quotient of the space $M(S)$ of Beltrami forms on S by the normal subgroup $\mathrm{QC}^0(S, \Sigma)$ of the group of quasiconformal homeomorphisms $\mathrm{QC}(S, \Sigma)$, consisting of the elements homotopic to the identity. The quotient map

$$M(S) \to M(S)/\,\mathrm{QC}^0(S, \Sigma) = \mathcal{T}(S, \Sigma)$$

induces a complex structure on the Teichmüller space, which makes it a complex manifold of dimension $3g-3+n$. Also, by Teichmüller's Theorem $\mathcal{T}(S, \Sigma)$ is homeomorphic to a ball in $\mathbb{R}^{6g-6+2n}$.

Using the marking by quasiconformal homeomorphisms, we define the Teichmüller metric between $x = (X, m)$ and $y = (Y, n)$ by

$$d_{\mathcal{T}}(x, y) = \frac{1}{2} \inf_{f \simeq n \circ m^{-1}} \{\log K(f) \mid f : X \to Y \ \text{ quasi-conformal}\},$$

where $K(f)$ is the quasiconformal dilatation, i. e.

$$K(f) = \frac{1 + \|\mu\|_\infty}{1 - \|\mu\|_\infty} \quad \text{with} \quad \mu = \frac{\frac{\partial f}{\partial \bar{z}}}{\frac{\partial f}{\partial z}}.$$

(These are distributional derivatives.) $d_{\mathcal{T}}$ turns $\mathcal{T}(S, \Sigma)$ into a complete metric space.

On $\mathcal{T}(S,\Sigma)$, we still have the left action[1] of the *mapping class group*

$$\Gamma(S,\Sigma) = \Gamma_{g,n} = \mathrm{Diffeo}^+(S,\Sigma)/\,\mathrm{Diffeo}^0(S,\Sigma) \cong \mathrm{QC}(S,\Sigma)/\,\mathrm{QC}^0(S,\Sigma).$$

Explicitly, for $f \in \Gamma(S,\Sigma)$ and $x = (X,m) \in \mathcal{T}(S,\Sigma)$,

$$f \cdot x = (X, m \circ f^{-1}).$$

$\Gamma(S,\Sigma)$ acts properly discontinuously by isometries for the Teichmüller metric. The quotient is the *moduli space of compact Riemann surfaces*

$$\mathcal{M}_{g,n} = \mathcal{T}(S,\Sigma)/\,\Gamma(S,\Sigma).$$

$\mathcal{M}_{g,n}$ is a complex orbifold of dimension $3g-3+n$. The Teichmüller metric descends to $\mathcal{M}_{g,n}$.

Looking at it from the point of view of algebraic geometry, $\mathcal{M}_{g,n}$ can also be viewed as a moduli stack. We will usually care about the coarse moduli space and denote this one by $\mathcal{M}_{g,n}$; it is a quasi-projective variety over $\mathbb{C}$. If a fine moduli space is needed, we will replace $\mathcal{M}_{g,n}$ with a finite cover $\mathcal{M}^{[\ell]}_{g,n}$ given by a level-structure (see Chapter 5). There is a good compactification $\overline{\mathcal{M}}_{g,n}$ of $\mathcal{M}_{g,n}$ due to P. Deligne and D. Mumford by adding stable curves [DM69]. Recall that a *stable curve* $\mathcal{C}$ over $\mathbb{C}$ is a connected algebraic curve, which has only ordinary double points as singularities, and every component of the normalization of $\mathcal{C}$ has negative Euler characteristic.

The cotangent space to a point $x = (X,m)$ in Teichmüller space $\mathcal{T}(S)$ can be naturally identified with the $\mathbb{C}$-vector space $\mathcal{Q}(X) = (\Omega^1_X)^{\otimes 2}(X)$ of holomorphic quadratic differentials. Since every holomorphic quadratic differential on X becomes the square of a holomorphic 1-form ω on a suitable double cover of X, it is convenient to consider instead the bundle $\Omega\,\mathcal{T}(S)$, whose fiber over $x = (X,m)$ is $\Omega^1_X(X) \setminus \{0\}$. Its points are in a natural way (marked) translation surfaces, and we will turn towards them in the next section.

[1] It is more naturally a right action. But since we want to consider $\Gamma_{g,n}$ as a fundamental group, we prefer to consider the left action. In remembrance of this fact, we use the right-quotient notation instead.

2.2 Translation Surfaces

Let X be an oriented, connected 2-dimensional manifold, carrying a translation atlas $A = \{(U_i, \varphi_i)\}_i$, i. e. the transition maps $\varphi_i \circ \varphi_j^{-1}$ are locally translations of $\mathbb{R}^2$. The datum (X, A) is called a *translation surface*. The case we will mostly care about, is when X can be embedded in a compact surface $\overline{X}$ with $\overline{X} \setminus X$ finite. Such surfaces arise from the following construction.

Let ω be a nonzero holomorphic 1-form on a compact Riemann surface X. We can define a translation atlas on $X \setminus Z(\omega)$, where $Z(\omega)$ is the set of zeros of ω, by using local primitives of ω as charts. The translation surface thus constructed will be denoted (X, ω). (Note that we secretly also remember the underlying compact surface X.)

We always identify $\mathbb{C}$ with $\mathbb{R}^2$ by sending $\{1, i\}$ to the standard basis. With this choice, a translation altas $\{(U_i, \varphi_i)\}_i$ on a surface leads in a natural way to a complex structure, since the multiplication by i on a chart leads to a well-defined complex structure on the tangent bundle. If we started from a compact Riemann surface X with non-zero holomorphic 1-form ω, then the complex structure induced by the translation structure is the original one.

On $X \setminus Z(\omega)$, one defines a flat Riemannian metric with trivial linear holonomy by pulling back the Euclidean metric via the coordinate charts. Geodesics for this metric are straight line segments; geodesics that connect two singularities are called *saddle connections*. The *group of relative periods* $\mathrm{Per}(\omega)$ is the subgroup of $\mathbb{R}^2$ spanned by the vectors $\int_\gamma \omega$ corresponding to saddle connections and loops γ. Alternatively, $\mathrm{Per}(\omega)$ is the image of $H_1(X, Z(\omega), \mathbb{Z})$, the homology relative to $Z(\omega)$, under integration over ω. A point $P \in Z(\omega)$ leads to a singularity of the translation structure, respectively of the metric: It is a conical point with a cone angle of $2\pi(d+1)$, where d is the multiplicity of the zero. By the Riemann-Roch theorem, ω has precisely $2g-2$ zeros counted with multiplicities. The volume form of the Riemannian metric is given by $\frac{i}{2}\omega \wedge \overline{\omega}$.

By a *translation covering*, we shall understand a non-constant holomorphic map $f : (X, \omega) \rightarrow (Y, \nu)$ between translation surfaces, such that $f^*\nu = \omega$.

The Moduli Space of Abelian Differentials

There is a natural stratification of $\Omega\mathcal{T}_g$ by the multiplicities of the zeros of the 1-form ω: Let $\kappa = (\kappa_1, \ldots, \kappa_{\ell(\kappa)})$ be a partition of $2g-2$. Then we denote by $\Omega\mathcal{T}_g(\kappa)$ the stratum of triples (X, f, ω), where ω has $\ell(\kappa)$ zeros $Z(\omega) = \{p_1, \ldots, p_{\ell(\kappa)}\}$ with p_i having multiplicity κ_i. We emphasize that there is no ordering on the set of zeros. An expression k^l indicates that there are l different zeros, all with the same multiplicity k.

Let $g \geq 2$ and let κ be a partition of $2g-2$. The stratum $\Omega\mathcal{T}_g(\kappa)$ is locally modeled on a cohomology space: Charts are provided by the periods of the 1-form. We briefly sketch the construction.To make this work, we have to refine the marking of points in $\Omega\mathcal{T}_g$. Let S be as in Section 2.1, and let $\Sigma \subset S$ be a finite set of $\ell(\kappa)$ points. Consider the following finite cover $\Omega\mathcal{T}(S)(\kappa)'$ of $\Omega\mathcal{T}(S)(\kappa)$: Let it consist of triples (X, m, ω), where $m : S \to X$ maps Σ to $Z(\omega)$. Say that two triples are equivalent if they only differ by an element of $\mathrm{QC}^0(S, [\Sigma])$. This is the subgroup of elements in $\mathrm{QC}^0(S)$, fixing Σ as a set. Choose a symplectic basis $\{a_i, b_i\}_{i=1}^{g}$ of $H_1(S, \mathbb{Z})$, and extend it to a basis of $H_1(S, \Sigma, \mathbb{Z})$ by $\ell(\kappa) - 1$ homotopy classes $\{c_i\}_{i=1}^{\ell(\kappa)-1}$ of paths γ_i, where γ_i connects p_i to p_{i+1}.

Let $x = (X, m, \omega) \in \Omega\mathcal{T}(S)(\kappa)'$. Integration of ω along the images of the symplectic basis yields a well-defined vector $\Phi(x) \in \mathbb{C}^{2g+\ell(\kappa)-1}$. Veech [Vee90] showed that this produces a local homeomorphism from $\Omega\mathcal{T}(S)(\kappa)'$ to $\mathbb{C}^{2g+\ell(\kappa)-1}$, whose coordinate changes are complex affine maps; they come from the change of the symplectic basis. Since they have determinant 1, we can locally pull back the Lebesgue measure on $\mathbb{C}^{2g+\ell(\kappa)-1}$, normalized such that the quotient torus $\mathbb{C}^{2g+\ell(\kappa)-1} / \mathbb{Z}[i]^{2g+\ell(\kappa)-1}$ has volume 1, and obtain a measure ν on the stratum.

If we factor out the action of the mapping class group on $\Omega\mathcal{T}_g$, we arrive at the moduli space $\Omega\mathcal{M}_g$ of isomorphism classes of pairs (X a compact Riemann surface, ω a non-zero holomorphic 1-form). The stratification carries over; the strata are denoted analogously by $\Omega\mathcal{M}_g(\kappa)$.

Each stratum of $\Omega\mathcal{M}_g$ decomposes into at most 3 connected components, who are distinguished by the properties "hyperelliptic", "non-hyperelliptic", "even" and "odd". We refer to [KZ03], and we will freely make use of the notation introduced therein to specify connected components of strata.

We have a canonical $\mathrm{GL}_2^+(\mathbb{R})$-action on $\Omega\,\mathcal{T}_g$, which preserves the stratification. For $A \in \mathrm{SL}_2(\mathbb{R})$ and $x = (X, m, \omega)$ a point in $\Omega\,\mathcal{T}_g$, let $A\cdot(X, m, \omega)$ denote the translation surface, obtained by postcomposing each chart $\varphi : U \to \mathbb{C}$ with the real affine map

$$z = x + iy \mapsto (1, i)A(x, y)^T.$$

The local $\mathrm{d}z$'s on the new coordinate charts glue together, and produce a holomorphic 1-form, denoted by $A \cdot \omega$. Hence, this defines a point $A \cdot x$ in $\Omega\,\mathcal{T}_g$. This action commutes with the action of Γ_g, thus it descends to an action on $\Omega\,\mathcal{M}_g$. Most of the time, we will consider the action of $\mathrm{SL}_2(\mathbb{R})$ or of the diagonal subgroup $\{g_t = \mathrm{diag}(e^t, e^{-t})\}$. The latter is called *Teichmüller geodesic flow.* Note also that the identity $X \to X$ gives an (orientation-preserving) homeomorphism between (X, m, ω) and $A \cdot (X, m, \omega)$, which we will denote by φ_A.

Assigning to $(X, \omega) \in \Omega\,\mathcal{M}_g$ its total area

$$\mathrm{Area}(X, \omega) = \tfrac{i}{2} \int_X \omega \wedge \overline{\omega}$$

defines a function $\Omega\,\mathcal{M}_g \to \mathbb{R}_{>0}$. Its level sets are preserved by the $\mathrm{SL}_2(\mathbb{R})$-action. It is sometimes convenient (see Section 8.2) to look only at a fixed level set; therefore, introduce the subspaces $\Omega_1\,\mathcal{T}_g$ and $\Omega_1\,\mathcal{M}_g$, where the 1-form has total area normalized to 1.

2.3 Teichmüller Curves

We summarize how a compact Riemann surface with a non-zero holomorphic quadratic differential q gives rise to a complex geodesic $\mathbb{H} \to \mathcal{M}_g$ in moduli space, and under which condition the image is an algebraic curve – called a Teichmüller curve. We confine ourselves to the case $q = \omega^{\otimes 2}$, a square of a non-zero holomorphic 1-form ω. Good references are e. g. [McM03], [Vee89], [HS06].

Affine Homeomorphisms

Given translation surfaces (X, ω), (Y, ν), we say that a homeomorphism $f : X \to Y$ is *affine*, if in local coordinates of the translation structures, f is given by

$$z \mapsto A \cdot z + t$$

for some $A \in \mathrm{GL}_2(\mathbb{R})$ and $t \in \mathbb{R}^2$. If f is affine, then its matrix part A is globally the same. If f is orientation preserving, then $\det(A) > 0$. The *affine group* $\mathrm{Aff}(X, \omega)$ is the group of all affine, orientation preserving homeomorphisms of X. Since X has finite volume, the matrix part of $f \in \mathrm{Aff}(X, \omega)$ is in $\mathrm{SL}_2(\mathbb{R})$. "Affine" can also be characterized in the following way: A homeomorphism $f : X \to Y$ is affine, if and only if it maps zeros to zeros, is smooth outside the zeros, and the pullback of the subspace $\mathrm{span}\{\mathrm{Re}\,\nu, \mathrm{Im}\,\nu\} \subset H^1(Y, \mathbb{R})$ by f is equal to $\mathrm{span}\{\mathrm{Re}\,\omega, \mathrm{Im}\,\omega\} \subset H^1(X, \mathbb{R})$.

Assigning to $f \in \mathrm{Aff}(X, \omega)$ its matrix part defines a group homomorphism

$$\mathrm{D} : \mathrm{Aff}(X, \omega) \to \mathrm{SL}_2(\mathbb{R}).$$

The image of D is called the *Veech group* of (X, ω), and is denoted by $\Gamma(X, \omega)$. Its kernel is the *group of translations* $\mathrm{Trans}(X, \omega)$, i. e. maps that are automorphisms for the translation structure. Finally, we call $\mathrm{Aut}(X, \omega) = \mathrm{D}^{-1}(\{\pm I\})$ the *group of affine biholomorphisms* of (X, ω). Note that $D(\mathrm{Aut}(X, \omega))$ is central in $\Gamma(X, \omega)$.

Veech [Vee89] showed that $\Gamma(X, \omega)$ is always a discrete subgroup of $\mathrm{SL}_2(\mathbb{R})$, i. e. a Fuchsian group. In general however, the quotient $\Gamma(X, \omega)\backslash\mathbb{H}$ has infinite volume. A translation surface (X, ω) is called *Veech surface*, if $\Gamma(X, \omega)$ is a lattice. Note also that $\Gamma(X, \omega)$ is necessarily non-uniform [Vee89], i. e. $\Gamma(X, \omega)\backslash\mathbb{H}$ is never compact.

A translation cover $p : (X, \omega) \to (Y, \nu)$ between Veech surfaces is called *Veech cover*, if the affine group of $(Y \setminus B, \nu)$ has finite index in $\mathrm{Aff}(Y, \nu)$, where B is the branch locus of p.

Teichmüller Disks

A *Teichmüller embedding* is a holomorphic embedding $\tilde{\jmath} : \mathbb{H} \to \mathcal{T}_g$, isometric w. r. t. the Poincaré metric on $\mathbb{H}$ and the Teichmüller metric on $\mathcal{T}_g$. A translation surface (X, ω) gives rise to a Teichmüller embedding in the following way: Send $A \in \mathrm{SL}_2(\mathbb{R})$ to $A \cdot (X, \mathrm{id}) = (A \cdot X, \varphi_A) \in \mathcal{T}(X)$, where $A \cdot X$ is the underlying complex structure of the translation structure of $A \cdot (X, \omega)$ and φ_A is as in Section 2.2. This map factors via $\mathrm{SO}(2) \backslash \mathrm{SL}_2(\mathbb{R})$ and yields a Teichmüller embedding

$$\tilde{\jmath} : \mathbb{H} \to \mathcal{T}(X),$$

where $\mathrm{SO}(2) \backslash \mathrm{SL}_2(\mathbb{R})$ is identified with $\mathbb{H}$ by $\mathrm{SO}(2)A \mapsto -\overline{A^{-1}(i)}$. The image $\tilde{\jmath}(\mathbb{H}) = \Delta = \Delta(X, \omega^{\otimes 2})$ is called a *Teichmüller disk*. It is a complex geodesic in $\mathcal{T}(X)$, which corresponds to the base point X and the cotangent vector $\omega^{\otimes 2}$. The affine group acts from the left on Δ as a subgroup of the mapping class group; in fact, it is the stabilizer of Δ in Γ_g [EG97, Lemma 5.2, Theorem 1]. It also acts from the left on $\mathrm{SO}(2) \backslash \mathrm{SL}_2(\mathbb{R})$ by $(B, \mathrm{SO}(2)A) \mapsto \mathrm{SO}(2)AB^{-1}$, and $\tilde{\jmath}$ is equivariant with respect to the two actions. Note that the action on $\mathbb{H}$ is not by Möbius transformations.

Remark 2.1 ([McM03], [HS06])
Let $B \in \mathrm{SL}_2(\mathbb{R})$. If $\mathbb{H}$ is identified with $\mathrm{SO}(2) \backslash \mathrm{SL}_2(\mathbb{R})$ as above, then the left action of B on $\mathrm{SO}(2) \backslash \mathrm{SL}_2(\mathbb{R})$ by $\mathrm{SO}(2)A \mapsto \mathrm{SO}(2)AB^{-1}$ corresponds to the left action of RBR on $\mathbb{H}$ by Möbius transformations, where $R : \mathbb{C} \to \mathbb{C}$, $z \mapsto -\overline{z}$.

The action of $\mathrm{Aff}(X, \omega)$ on Δ need not be effective. The pointwise stabilizer $\mathrm{Stab}_0(\tilde{\jmath}(\mathbb{H}))$ is isomorphic to $\mathrm{Aut}(X, \omega) = \mathrm{D}^{-1}(\{\pm I\})$; this is the group of biholomorphisms of X that propagate to the entire Teichmüller disk $\Delta(X, \omega^2)$.

Passing to the quotient by $\mathrm{Aff}(X, \omega)$, respectively by Γ_g, on both sides of $\tilde{\jmath} : \mathbb{H} \to \mathcal{T}_g$, we get a holomorphic isometric immersion $j : \mathbb{H} / \mathrm{Aff}(X, \omega) = \mathbb{H} / \Gamma(X, \omega) \to \mathcal{M}_g$. In the general case, the image of j in $\mathcal{M}_g$ will be something wild; however,

Proposition 2.2 ([McM03, Corollary 3.3])
The map $j : \mathbb{H}/\Gamma(X,\omega) \to \mathcal{M}_g$ covers an algebraic curve $\mathcal{C}$ in the moduli space, if and only if $\Gamma(X,\omega)$ is a lattice in $\mathrm{SL}_2(\mathbb{R})$. *In this case, j is generically injective and $\mathbb{H}/\Gamma(X,\omega)$ is the normalization of $\mathcal{C}$. j is then called a* Teichmüller curve.

The whole discussion can be lifted to the "tangent bundles" $\Omega\mathcal{T}(X)$, respectively $\Omega\mathcal{M}_g$. Then we are considering the $\mathrm{SL}_2(\mathbb{R})$-orbit of (X, id, ω), respectively (X,ω). In this case,

Proposition 2.3 ([SW04, Proposition 8])
The $\mathrm{SL}_2(\mathbb{R})$*-orbit of (X,ω) is closed in $\Omega\mathcal{M}_g$, if and only if (X,ω) is a Veech surface.*

2.4 Origamis

In this section, we formalize the definition of an origami, and list (and prove) some general facts about origamis. I have first met them through [Sch05a], and I should also like to mention [Kre10] as a good reference for the combinatorial aspects.

Definition 2.4 Let $E = \mathbb{C}/(\mathbb{Z} \oplus i\,\mathbb{Z})$ be the square torus. An *origami* $\mathbf{O}$ is a compact Riemann surface, together with a non-constant holomorphic map $\pi : \mathbf{O} \to E$ which is ramified at most over $\infty := \overline{0} \in E$.

The $\mathbb{C}$-vector space $\Omega^1_E(E)$ is 1-dimensional. So a non-zero holomorphic 1-form on E is unique up to multiplication by $\mathbb{C}^\times$. If not stated otherwise, we will assume that it has been chosen as the 1-form ω_E induced by $\mathrm{d}z$ on $\mathbb{C}$ via the universal covering $\mathbb{C} \to E$, so that integration $E \to \mathbb{C}$, $x \mapsto \int_\infty^x \omega_E \bmod \mathrm{Per}(\omega_E)$ produces the isomorphism from the uniformization.

Remark 2.5 Let $\pi : \mathbf{O} \to E$ be an origami. The monodromy of π induces a description of $\mathbf{O}$ as the result of gluing of finitely many unit squares. Therefore, some authors prefer the name "square-tiled surface". However, be aware that this term is also used for pillowcase covers (see Remark 2.19

below), which are, though closely related to origamis, not the same. The name "origami" was invented by P. Lochak [Loc05], who studied them in the context of the action of $\mathrm{Gal}(\overline{\mathbb{Q}}/\mathbb{Q})$ in Grothendieck-Teichmüller theory.

Remark 2.6 An origami defines a Veech surface by pulling back the non-zero holomorphic 1-form ω_E on E. This is a general phenomenon: If (Y,ν) is a Veech surface and $p : X \to Y$ is a holomorphic covering map, then $(X, p^*\nu)$ is again a Veech surface, provided that the Veech group of $(Y \setminus B, \nu)$ is a lattice. Here, B is the branch locus of p. Indeed, by [GJ00, Theorem 4.9], the Veech groups $\Gamma(X, p^*\nu)$ and $\Gamma(Y,\nu)$ are commensurate. Veech surfaces arising in this way are called *geometrically imprimitive*, and else *geometrically primitive.*

Note that there is also the stronger notion of *algebraic primitivity* of a Veech surface: A genus g Veech surface (X,ω) is algebraically primitive, if its trace field

$$K(X,\omega) = \mathbb{Q}(\{\mathrm{Tr}(\gamma) \mid \gamma \in \Gamma(X,\omega)\})$$

has degree g over $\mathbb{Q}$. (Note that $K(X,\omega)$ is always a number field of degree $\leq g$ by [McM03, Theorem 5.1]). An algebraically primitive Veech surface is geometrically primitive, but the converse is false in general [McM06].

In the following, we will consider an origami $\mathbf{O}$ as being endowed with the translation structure from Remark 2.6. We write $\mathrm{Aff}(\mathbf{O})$ for the affine group, $\Gamma(\mathbf{O})$ for the Veech group, $\mathrm{Aut}(\mathbf{O})$ for the group of affine biholomorphisms and $\mathrm{Trans}(\mathbf{O})$ for the group of translations of $\mathbf{O}$. By virtue of Proposition 2.2, an origami $\mathbf{O}$ of genus g also defines a Teichmüller curve, called *origami curve*, which we denote by $j(\mathbf{O}) : \mathcal{C}(\mathbf{O}) = \mathbb{H}/\Gamma(\mathbf{O}) \to \mathcal{M}_g$.

Remark 2.7 Let $\pi : \mathbf{O} \to E$ be an origami. The Teichmüller curve comprises all points in $\mathcal{M}_g$, which can be reached by an affine shear $A \cdot \mathbf{O}$ with $A \in \mathrm{SL}_2(\mathbb{R})$. This can also be seen as varying the translation structure on E by A, and then pulling back $A \cdot \omega$ on $A \cdot E$ via $\varphi_A \circ \pi$; the surface thus constructed is isomorphic as a translation surface to $A \cdot \mathbf{O}$. Thus, the family of curves $\{A \cdot \mathbf{O}\}_{A \in \mathrm{SL}_2(\mathbb{R})}$ arises by variation of the elliptic curve on the base, i. e. by variation of the τ-invariant.

Definition 2.8 A Veech surface (X, ω) is called *arithmetic*, if it admits a translation cover $\pi : X \to T$ of a torus, ramified over at most one point.

By the preceding remark, a Veech surface is arithmetic, if and only if it is (isomorphic to) a point in the $\mathrm{SL}_2(\mathbb{R})$-orbit of an origami (up to scaling of the holomorphic 1-form). A characterization of arithmetic Veech surfaces is given in [GJ00, Theorem 5.5].

Remark 2.9 a) A translation surface (X, ω) is an origami, i. e. admits a translation covering $p : X \to E$, ramified over one point, if and only if the lattice of relative periods $\mathrm{Per}(\omega)$ is a subgroup of $\mathbb{Z} \oplus i\,\mathbb{Z}$ (see [Kre10, Proposition 1.8] for a proof of this fact). An origami is called *primitive* (not to be confused with the above mentioned notion of primitivity), if $\mathrm{Per}(\omega) = \mathbb{Z} \oplus i\,\mathbb{Z}$.

b) An origami is primitive, if and only if it does not come from subdividing the squares of another origami, i. e. if and only if there is no translation covering $\pi' : \mathbf{O} \to E'$ to a torus E', ramified over at most one point, such that π factors over π' as $\pi = f \circ \pi'$ with a translation cover $f : E' \to E$, such that $\deg(f) > 1$.

c) If $\pi : \mathbf{O} \to E$ is a primitive origami of genus $g \geq 2$, then $\mathrm{Aff}(\mathbf{O}^*) = \mathrm{Aff}(\mathbf{O})$, where $\mathrm{Aff}(\mathbf{O}^*)$ is the subgroup of $\mathrm{Aff}(\mathbf{O})$ fixing the fiber $\pi^{-1}(\infty)$ as a set.

Proof: b) Assume that there exists a $\pi' : \mathbf{O} \to E'$, such that $\pi = f \circ \pi'$. Necessarily, π' is, if at all, ramified over a point $p \in f^{-1}(\infty)$. We have a commutative diagram

$$\begin{array}{ccc} E' & \xrightarrow{\quad f \quad} & E \\ \big\downarrow & & \big\downarrow \\ \mathbb{C}\,/\,\mathrm{Per}(f^*\omega_E) & \longrightarrow & \mathbb{C}\,/\,\mathrm{Per}(\omega_E) \end{array}$$

where the vertical arrows are the isomorphisms $\int_p^{\cdot} f^*\omega_E$, respectively $\int_\infty^{\cdot} \omega_E$ given by integration, which is well-defined modulo the period lattices $\mathrm{Per}(f^*\omega_E)$ and $\mathrm{Per}(\omega_E)$, and the bottom horizontal arrow is the projection induced by the inclusion of $\mathrm{Per}(f^*\omega_E)$ in $\mathrm{Per}(\omega_E)$. The diagram commutes, for

$$\int_{f(p)}^{f(x)} \omega_E \equiv \int_p^x f^*\omega_E \bmod \mathrm{Per}(\omega_E).$$

Let γ be a closed loop or a saddle connection in $\mathbf{O}$. Since π' is ramified at most over one point, $\pi' \circ \gamma$ is a closed loop on E'. Now,

$$\int_\gamma \omega = \int_\gamma \pi'^*(f^*\omega_E) = \int_{\pi'\circ\gamma} f^*\omega_E \in \mathrm{Per}(f^*\omega_E) \subsetneq \mathrm{Per}(\omega_E),$$

since f has at least degree 2.

Conversely, assume that $\mathbf{O}$ is not primitive. From part a), we know $\mathrm{Per}(\omega) \leqslant \mathbb{Z} \oplus i\,\mathbb{Z}$; by assumption, this inclusion is strict. Consider the torus $E' = \mathbb{C}\,/\,\mathrm{Per}(\omega)$, and the map $\pi' : \mathbf{O} \to E'$, induced by integration $x \mapsto \int_q^x \omega$, where $q \in \pi^{-1}(\infty)$ shall be a conical point, if there are any (or arbitrary if not). π' is holomorphic and ramified at most over $\overline{0}$, since every conical point of $\mathbf{O}$ maps to $\overline{0}$. The composition of π' with the projection $\mathbb{C}\,/\,\mathrm{Per}(\omega) \to \mathbb{C}\,/(\mathbb{Z} \oplus i\,\mathbb{Z})$ and the uniformization isomorphism $\mathbb{C}\,/(\mathbb{Z} \oplus i\,\mathbb{Z}) \cong E$ produces a factorization of π.

c) We have to show that $\pi(f(p)) = \pi(p)$ for any point p in $\pi^{-1}(\infty)$ and any $f \in \mathrm{Aff}(\mathbf{O})$. Since $g \geq 2$, there is at least one conical point $q \in \mathbf{O}$. (Otherwise, $\mathbf{O}$ would admit a complete Riemannian metric of constant curvature 0, contradicting $g \geq 2$.) Any $f \in \mathrm{Aff}(\mathbf{O})$ maps q to a conical point $f(q)$. Therefore it acts on the saddle connections, and also on the vectors in $\mathbb{C}$ associated with saddle connections. Explicitly, this action is given by the rule

$$f \cdot \Big(\int_\gamma \omega\Big) = \int_{f\circ\gamma} \omega = \int_\gamma f^*\omega = A \cdot \Big(\int_\gamma \omega\Big)\,,$$

where $A = \mathrm{D}(f)$ acts as affine transformation of $\mathbb{R}^2 = \mathbb{C}$. So A preserves $\mathrm{Per}(\omega) = \mathbb{Z} \oplus i\,\mathbb{Z}$, i. e. $A \in \mathrm{SL}_2(\mathbb{Z})$. Take a path γ_1 joining p to q, and a path γ_2 joining q to $f(q)$ (e. g. a saddle connection). Then the integrals $\int_{\gamma_i} \omega$ are in $\mathbb{Z} \oplus i\,\mathbb{Z}$, and so is $\int_{f\circ\gamma_1} \omega$. Therefore

$$\int_p^{f(p)} \omega \equiv \int_{f\circ\gamma_1^{-1}} \omega + \int_{\gamma_2} \omega + \int_{\gamma_1} \omega \equiv 0 \bmod \mathbb{Z} \oplus i\,\mathbb{Z}\,.$$

This shows that $x \mapsto \int_p^x \omega$ maps $f(p)$ to $\overline{0} = \infty$. ■

Combinatorics

The starting point for the combinatorial discussion of origamis is the identification of the fundamental group of the once-punctured torus with F_2. Let again $E = \mathbb{C}/(\mathbb{Z} \oplus i\,\mathbb{Z})$, and let $\infty := \overline{0} \in E$. We set $E^* = E \setminus \{\infty\}$. Moreover, we choose a base point $e \in E^*$. (w. l. o. g. no coordinate of e is integral.) As generators of $\pi_1(E^*, e)$, we choose x, the (homotopy class of the) horizontal loop based at e and y, the (homotopy class of the) vertical loop based at e, and fix the isomorphism $\pi_1(E^*, e) \to F(\{x, y\}) = F_2$ induced by this choice.

Proposition 2.10 *An origami $\pi : \mathbf{O} \to E$ of degree d is equally well given by*

- *a conjugacy class of a subgroup $H = H(\mathbf{O}) \leqslant F_2$ of index d,*
- *two permutations σ_x, $\sigma_y \in S_d$, which generate a transitive subgroup (up to the action of $\mathrm{Inn}(S_d)$)*

Proof: We sketch how to pass from one description to another. A complete discussion can be found in [Kre10, Sect. 1.1].

To obtain a conjugacy class of a subgroup in F_2, do the following. Let $\mathbf{O}^* = \mathbf{O} \setminus \pi^{-1}(\infty)$. Choose a base point $x \in \pi^{-1}(e)$. Then $\pi_1(\mathbf{O}^*, x)$ injects into $\pi_1(E^*, e)$, because $\pi : \mathbf{O}^* \to E^*$ is a topological covering map. If we choose another base point x', then the image changes by conjugation $H \mapsto wHw^{-1}$, where w is the image of a path connecting x to x' in $\mathbf{O}^*$. Conversely, any conjugacy class of a subgroup determines a covering, unique up to fiber-preserving homeomorphism (see [Hat02, Theorem 1.38]).

To obtain two permutations from $\pi : \mathbf{O} \to E$, consider the monodromy action of $\pi_1(E^*, e)$ on the fiber $\pi^{-1}(e)$. This defines a group homomorphism $\sigma : \pi_1(E^*, e) \to S_d$, unique up to renumbering. Set $\sigma_x = \sigma(x)$ and $\sigma_y = \sigma(y)$. Together they generate a transitive subgroup of S_d, since $\mathbf{O}$ is connected. Conversely, given σ_x, $\sigma_y \in S_d$, which together act transitively, take d unit squares and glue the right side of square i to the left side of square $\sigma_x(i)$ and the upper side of square i to the lower side of square $\sigma_y(i)$. This defines an origami of degree d.

To obtain a conjugacy class of a subgroup of F_2 from two permutations σ_x, σ_y, generating a transitive subgroup, consider the group homomorphism

$$\rho : F_2 \to S_d\ , \quad x \mapsto \sigma_x, \quad y \mapsto \sigma_y.$$

Since $\rho(F_2)$ is transitive, $\mathrm{Stab}_{\rho(F_2)}(1)$ is a subgroup of index d in $\rho(F_2)$, and $H = \rho^{-1}(\mathrm{Stab}_{\rho(F_2)}(1))$ has index d in F_2. Choosing $m \in \{1, \ldots, d\}$ instead of 1 replaces H by a conjugate.

To obtain two permutations from a conjugacy class of a subgroup $H \leqslant F_2$ of index d, consider the action of F_2 on the left cosets $\{a_1 \cdot H, a_2 \cdot H, \ldots, a_d \cdot H\}$ of H in F_2 by left multiplication. This defines two permutations σ_x, $\sigma_y \in \mathrm{Sym}(F_2/H) \cong S_d$, which clearly generate a transitive subgroup. ∎

Remark 2.11 Let $\pi : \mathbf{O} \to E$ be an origami of degree d, and let $\mathbf{O}^* = \mathbf{O} \setminus \pi^{-1}(\infty)$. Then $\mathbf{O}^*$ is homotopy equivalent to a 4-valent graph $\mathcal{G}(\mathbf{O})$ with d vertices.

PROOF: $\mathbf{O}$ is a surface tiled by squares. Let $\mathcal{G}(\mathbf{O})$ be the dual graph of the tiling. This is clearly a deformation retract of $\mathbf{O}^*$. ∎

Proposition 2.10 allows for a description of the Veech group as a stabilizer in the automorphism group of the free group $F_2 = F(\{x, y\})$. This was first observed by G. Weitze-Schmithüsen [Sch05a], [Sch04]. Recall that $\mathrm{Out}(F_2) = \mathrm{Aut}(F_2)/\mathrm{Inn}(F_2)$ is isomorphic to $\mathrm{GL}_2(\mathbb{Z})$: Consider the homomorphism $\beta : \mathrm{Aut}(F_2) \to \mathrm{GL}_2(\mathbb{Z})$ induced by the abelianization map $F_2 \to \mathbb{Z}^2$; β is given by

$$f \mapsto A = \begin{pmatrix} a & b \\ c & d \end{pmatrix} = \begin{pmatrix} \#_x(f(x)) & \#_x(f(y)) \\ \#_y(f(x)) & \#_y(f(y)) \end{pmatrix},$$

where $\#_x(w)$ is the number of letters x minus the number of letters x^{-1} in a word $w \in F_2$, and $\#_y$ is analogously defined. Then one can show that $\mathrm{Ker}(\beta) = \mathrm{Inn}(F_2)$. The subgroup $\mathrm{Aut}^+(F_2)$ is now defined as the preimage of $\mathrm{SL}_2(\mathbb{Z})$ under β.

Theorem 2.12 ([Sch04, Proposition 2.1])
Let $\mathbf{O}$ *be a primitive origami, and let* $H = H(\mathbf{O})$ *be a representative of the conjugacy class of subgroups of* F_2 *associated with* $\mathbf{O}$*. Define the stabilizer subgroup of* H *by*

$$\mathrm{Stab}_{\mathrm{Aut}^+(F_2)}(H) = \{f \in \mathrm{Aut}^+(F_2) \mid f(H) = H\}.$$

Then

$$\Gamma(\mathbf{O}) = \beta(\mathrm{Stab}_{\mathrm{Aut}^+(F_2)}(H)).$$

In particular, this implies that $\Gamma(\mathbf{O})$ has finite index in $\mathrm{SL}_2(\mathbb{Z})$: For $\mathrm{Stab}_{\mathrm{Aut}^+(F_2)}(H)$ has finite index in $\mathrm{Aut}^+(F_2)$ by virtue of being a stabilizer of the action of $\mathrm{Aut}^+(F_2)$ on the finite set of index d-subgroups of F_2.

We can also retrieve the group of translations in this description. Let $\pi : \mathbf{O} \to E$ be an origami, and let us continue to use the notations from Proposition 2.10. Fix a base point $x \in \pi^{-1}(e)$, so that we have a subgroup $H = H(\mathbf{O}) \leq \pi_1(E^*, e)$ associated with $\mathbf{O}$. We also fix a universal covering $u : \widetilde{X} \to \mathbf{O}^*$ (with base point $\tilde{x}$ lying over x), and endow $\widetilde{X}$ with the translation structure obtained from $\widetilde{\omega} = (\pi \circ u)^* \omega_E$. Now consider the fundamental groups as Galois groups acting on $\widetilde{X}$, i.e. $H = \mathrm{Gal}(\widetilde{X}/\mathbf{O}^*)$ and $F_2 = \mathrm{Gal}(\widetilde{X}/E^*)$; both groups lie in $\mathrm{Aff}(\widetilde{X}, \widetilde{\omega})$ (which, in the case of a primitive origami, is isomorphic to $\mathrm{Aut}^+(F_2)$ by [Sch04, Lemma 2.8]).

Proposition 2.13 *Let* $\pi : \mathbf{O} \to E$ *be an origami of genus* $g \geq 2$*, and* $H = H(\mathbf{O})$ *as above. Then* $\mathrm{Trans}(\mathbf{O}) \cong N(H)/H$*, where* $N(H)$ *is the normalizer of* H *in* $F_2 = \pi_1(E^*, e)$*.*

PROOF: First note that $\mathrm{Trans}(\mathbf{O}) = \mathrm{Trans}(\mathbf{O}^*)$, where the latter is the subgroup of translations preserving $\pi^{-1}(\infty)$. This follows from the fact that translations act trivially on the vectors associated with saddle connections and the proof of Remark 2.9. With the notations introduced above, let

$$\mathrm{Aff}_u(\widetilde{X}, \widetilde{\omega}) = \{f \in \mathrm{Aff}(\widetilde{X}, \widetilde{\omega}) \mid f \text{ descends to } \overline{f} \in \mathrm{Aff}(\mathbf{O}^*) \text{ via } u\},$$

and let $\mathrm{Trans}_u(\widetilde{X}, \widetilde{\omega}) = \mathrm{Aff}_u(\widetilde{X}, \widetilde{\omega}) \cap \mathrm{Gal}(\widetilde{X}/E^*)$.

Every element $\overline{f}$ in Trans($\mathbf{O}^*$) lifts to some $f \in \mathrm{Trans}_u(\widetilde{X}, \widetilde{\omega})$, which provides a surjective homomorphism

$$\mathrm{Trans}_u(\widetilde{X}, \widetilde{\omega}) \to \mathrm{Trans}(\mathbf{O}^*)\ , \quad f \mapsto \overline{f}$$

whose kernel is precisely $H = \mathrm{Gal}(\widetilde{X}/\,\mathbf{O}^*)$. So the claim follows if we show that the subgroup $\mathrm{Trans}_u(\widetilde{X}, \widetilde{\omega})$ of $\mathrm{Gal}(\widetilde{X}/E^*) = F_2$ is equal to $N(H)$. For let $g \in \mathrm{Trans}_u(\widetilde{X}, \widetilde{\omega})$, and let $h \in H$. Then h descends to $\overline{h} = \mathrm{id}_{\mathbf{O}^*}$. This implies $\overline{ghg^{-1}} = \mathrm{id}_{\mathbf{O}^*}$, so $ghg^{-1} \in H$. Conversely, for an element $g \in N(H)$ define $\overline{g} : \mathbf{O}^* \to \mathbf{O}^*$ by $H \cdot x \mapsto H \cdot g(x)$. Since for every $h \in H$, there exists $h' \in H$ such that $g(h(x)) = h'(g(x))$ for every $x \in \widetilde{X}$, this is well-defined. Thus $g \in \mathrm{Trans}_u(\widetilde{X}, \widetilde{\omega})$. ■

Definition 2.14 An origami $\pi : \mathbf{O} \to E$ is called *normal* (or Galois), if π is a normal covering map.

By standard covering space theory, an origami $\mathbf{O}$ is normal, if and only if $H(\mathbf{O})$ is a normal subgroup of $F_2 = \pi_1(E^*, e)$. Moreover, if $\mathbf{O}$ is an origami of genus ≥ 2 and degree d, then this is equivalent to $|\,\mathrm{Trans}(\mathbf{O})| = d$. A discussion of normal origamis can be found in [Kre10].

Finally, we remark that we can also describe the ramification of $\pi : \mathbf{O} \to E$ combinatorially.

Remark 2.15 Let $\pi : \mathbf{O} \to E$ be an origami of degree d, let σ_x and σ_y be the permutations associated with $\mathbf{O}$ by Proposition 2.10. The monodromy of the path $yxy^{-1}x^{-1}$ on E describes the ramification behavior above ∞: There is a bijection between the equivalence classes of lower left corners of the squares of $\mathbf{O}$ and the orbits of $k = \sigma_y\sigma_x\sigma_y^{-1}\sigma_x^{-1}$, given by assigning to the lower left corner of the square i the orbit of i. Each orbit corresponds thus to a cycle in the cycle decomposition of k. The ramification index of π at $p \in \pi^{-1}(\infty)$ is equal to the cycle length of the cycle associated with p. Recall that we determine the stratum of $\mathbf{O}$ from the multiplicities of the zeros of $\pi^*\omega_E$. A point $p \in \pi^{-1}(\infty)$ of ramification index e leads to a zero of order $e - 1$.

Cusps

Recall that a holomorphic quadratic differential q on a compact Riemann surface X is called *Strebel*, if the horizontal geodesic flow for the flat Riemannian metric induced by q is completely periodic. More generally, $\theta \in [0, 2\pi)$ is called a Strebel direction for q, if $e^{-i2\theta}q$ is Strebel.

If we specialize to a Veech surface (X, ω), then the Veech alternative (see e. g. [Vor96, Theorem 3.4]) asserts that the 1-form ω (or rather $\omega^{\otimes 2}$) is Strebel in direction θ, if and only if there exists a parabolic element $A \in \Gamma(X, \omega)$ with θ as an eigen-direction, i. e. $Av = v$ for any vector v in direction θ. The conjugacy classes of maximal parabolic subgroups in $\overline{\Gamma}(X, \omega)$ are in turn in bijection with the cusps of $\mathbb{H}/\Gamma(X, \omega)$ (or if you prefer, with the cusps of $R\Gamma(X, \omega)R\backslash\mathbb{H}$, see Remark 2.1). Here, $\overline{\Gamma}(X, \omega)$ denotes the image of $\Gamma(X, \omega)$ in $\mathrm{PSL}_2(\mathbb{R})$.

There is a stable curve corresponding to each of the cusps:

Proposition 2.16 *Let (X, ω) be a Veech surface, and let $\mathcal{C} = \mathbb{H}/\Gamma(X, \omega)$ be the associated Teichmüller curve. Let $\overline{\mathcal{C}}$ be the completion of $\mathcal{C}$ to a compact Riemann surface. Then*

a) the map $j : \mathcal{C} \to \mathcal{M}_g$ extends to a map $\overline{\jmath} : \overline{\mathcal{C}} \to \overline{\mathcal{M}}_g$

b) the points in $\mathrm{Im}(\overline{\jmath}) \cap \partial\mathcal{M}_g$ are given by stable curves, obtained by contracting the waist curves of all cylinders in the cylinder decomposition associated with Strebel directions of ω.

This was shown by Masur [Mas75] for the case of Teichmüller disks; an adaption to Teichmüller curves can be found in [HS06, Sect. 4]. There is a more precise formulation in the case of origamis. In the following, $T = \left(\begin{smallmatrix} 1 & 1 \\ 0 & 1 \end{smallmatrix}\right) \in \mathrm{SL}_2(\mathbb{Z})$.

Corollary 2.17 *Let $\pi : \mathbf{O} \to E$ be a primitive origami, and let $\omega = \pi^*\omega_E$. Then*

a) the T-orbits of the action of $\mathrm{SL}_2(\mathbb{Z})$ on $\mathrm{PSL}_2(\mathbb{Z})/\overline{\Gamma}(\mathbf{O})$ by left multiplication are in bijection with the cusps of $\mathcal{C}(\mathbf{O}) = \mathbb{H}/\Gamma(\mathbf{O})$.

b) the cusp corresponding to the T-orbit of $A\overline{\Gamma}(\mathbf{O}) \in \mathrm{PSL}_2(\mathbb{Z})/\overline{\Gamma}(\mathbf{O})$ is mapped to the (isomorphism class of the) stable curve in $\overline{\mathcal{M}}_g$, obtained by contracting the waist curves of the horizontal cylinders of $A \cdot \mathbf{O}$.

Proof: a) Let c be a cusp of $\mathcal{C}(\mathbf{O})$. By the above, c can be interpreted as a conjugacy class of a maximal parabolic subgroup of $\overline{\Gamma}(\mathbf{O})$. Pick a representative subgroup, and let P be a generator of it. Then P is conjugate in $\mathrm{PSL}_2(\mathbb{Z})$ to $\pm T^k$, i.e. $P = \pm A^{-1}T^kA$ with $A \in \mathrm{SL}_2(\mathbb{Z})$ and $k \in \mathbb{Z}$. We assign to c the T-orbit of $A\overline{\Gamma}(\mathbf{O})$. This is well-defined: If P' is a generator of another subgroup in the same conjugacy class as $\langle P\rangle$, then $P' = \pm B^{-1}P^{\varepsilon}B$ for some $B \in \Gamma(\mathbf{O})$ and $\varepsilon \in \{\pm 1\}$; thus, $P' = \pm(AB)^{-1}T^{\varepsilon\, k}AB$, and $AB\overline{\Gamma}(\mathbf{O}) = A\overline{\Gamma}(\mathbf{O})$. Now, we construct an inverse: let

$$A\overline{\Gamma}(\mathbf{O}), TA\overline{\Gamma}(\mathbf{O}), \ldots, T^{k-1}A\overline{\Gamma}(\mathbf{O})$$

be the T-orbit of $A\overline{\Gamma}(\mathbf{O})$ in $\mathrm{PSL}_2(\mathbb{Z})/\overline{\Gamma}(\mathbf{O})$ (with $k \in \mathbb{N}$). Then $P = \pm A^{-1}T^kA \in \overline{\Gamma}(\mathbf{O})$ is a parabolic element. Moreover, $\pm A^{-1}T^jA \notin \overline{\Gamma}(\mathbf{O})$ for $j < k$, so P generates a maximal parabolic subgroup in $\overline{\Gamma}(\mathbf{O})$.

b) Let $P = \pm A^{-1}T^kA$ be a generator of a maximal parabolic subgroup, obtained from $\langle T\rangle A\overline{\Gamma}(\mathbf{O})$ via the bijection in a). By the above, the 1-form ω is Strebel in direction $\theta = \theta(v)$, where v is an eigenvector of P. Shearing by A sends the direction v to Av, which is an eigenvector for T^k. Hence, the 1-form $A \cdot \omega$ on $A \cdot \mathbf{O}$ is Strebel for the horizontal direction, and we conclude with Proposition 2.16. ■

Origamis and Pillowcase Covers

Let us close this part by pointing out the connection between origamis and the also very popular pillowcase covers (see [Wri11], [FMZ10], [EKZ10a]). A pillowcase is the sphere $\mathbb{P}^1$ with four marked points $z_1, \ldots, z_4$, endowed with the holomorphic quadratic differential

$$q_0 = \frac{(\mathrm{d}z)^{\otimes 2}}{(z-z_1)(z-z_2)(z-z_3)(z-z_4)}.$$

The associated half-translation surface[2] is depicted in Figure 2.1. A pil-

[2] i. e. the transition maps of charts are locally $z \mapsto \pm z + c$, $c \in \mathbb{C}$

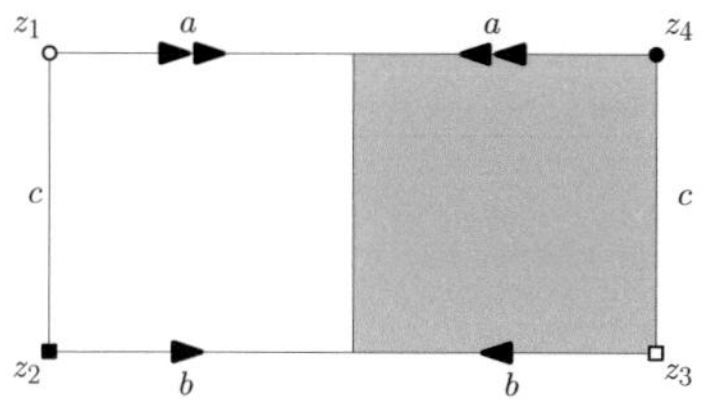

Figure 2.1: *A pillowcase. Sides with the same letters are glued.*

lowcase cover is a cover $X \to \mathbb{P}^1$, ramified at most over $z_1, \ldots, z_4$. Basic examples are cyclic covers. They arise from the desingularization of

$$y^N - (z - z_1)^{a_1}(z - z_2)^{a_2}(z - z_3)^{a_3}(z - z_4)^{a_4}$$

under the condition that

$$N > 1\ , \quad \gcd(N, a_1, \ldots, a_4) = 1\ , \quad 0 < a_i \leq N\ , \quad \sum_{i=1}^{4} a_i \equiv 0 \bmod N.$$

The resulting surface is denoted $M_N(a_1, \ldots, a_4)$. The most basic example is $N = 2$ and $a_i = 1$ for $i = 1, \ldots, 4$, in which case we obtain a cover $\pi_2 : E_2 \to \mathbb{P}^1$ from a torus with 4 marked points $e_1, \ldots, e_4$, which are 2-torision points (for any choice of $x \in \{e_1, \ldots, e_4\}$ as base point of the elliptic curve). Postcomposition with $[2] : E_2 \to E_2$, the multiplication by $[2]$ on (E, x) gives a translation cover of a torus, which maps $e_1, \ldots, e_4$ to the single point x.

If $\pi : X \to \mathbb{P}^1$ is a pillowcase cover, and $q = \pi^* q_0$ is not the square of a global holomorphic 1-form, then there is a canonical double cover $k : \hat{X} \to X$ such that $k^* q = \hat{\omega}^{\otimes 2}$ for some non-zero holomorphic 1-form $\hat{\omega}$ on $\hat{X}$. We have (see also [Wri11])

Remark 2.18 The differential q is not a square of a holomorphic 1-form on X, if and only if $\pi^*(z - z_1)(z - z_2)(z - z_3)(z - z_4)$ is not a square in the function field $\mathcal{M}(X)$ of X.

Proof: If $q = \omega \otimes \omega$ for some holomorphic 1-form ω on X, then in a chart (U, φ) of X, we have

$$q = F^{-1}(\mathrm{d}\pi)^{\otimes 2} = F^{-1}\Big(\frac{\partial \pi}{\partial \varphi}\Big)^2 (\mathrm{d}\varphi)^{\otimes 2},$$

where $F = (\pi - z_1)\cdots(\pi - z_4)$. Also, we can write $q = \omega \otimes \omega = g^2(\mathrm{d}\varphi)^{\otimes 2}$ with $g \in \mathcal{O}_X(U)$. Setting

$$G_\varphi = g^{-1}\frac{\partial \pi}{\partial \varphi},$$

we have locally $F = G_\varphi^2$, and the G_φ (for different charts) glue together to a holomorphic function on X. Therefore the image of $(z - z_1)\cdots(z - z_4)$ in $\mathcal{M}(X)$ is a square. Conversely, if $F = G^2$ is a square, then $\omega = \frac{\mathrm{d}\pi}{G}$ defines a holomorphic 1-form on X, whose square is q. For consider $\mathrm{ord}_P(\omega)$ for $P \in X$. If P is not a point lying over one of the points $z_1, \ldots, z_4$, then π and G are locally invertible, so $\mathrm{ord}_P(\omega) = 0$; otherwise,

$$2\,\mathrm{ord}_P(G) = \mathrm{ord}_P(F) = e_P(\pi) = k \geq 1,$$

so $\mathrm{ord}_P(G) = k' \geq 1$, with $2k' = k$. Also, locally around P, $\mathrm{d}\pi = \mathrm{d}z^k = kz^{k-1}\mathrm{d}z$, so

$$\mathrm{ord}_P(\omega) = \mathrm{ord}_P(\mathrm{d}\pi) - \mathrm{ord}_P(G) = 2k' - 1 - k' = k' - 1 \geq 0. \qquad \blacksquare$$

Remark 2.19 a) If $\pi : \mathbf{O} \to E$ is an origami, then postcomposing with the quotient $E \to \mathbb{P}^1$ by the involution $[-1]$ (for the elliptic curve (E, ∞)) produces a pillowcase cover.

b) If $\pi : X \to \mathbb{P}^1$ is a pillowcase cover, and q is a square of a holomorphic 1-form ω, then π factors over π_2, i. e. $\pi = \pi' \circ \pi_2$ for a covering map $\pi' : X \to E_2$, ramified over $e_1, \ldots, e_4$. After postcomposition with $[2]$, we see that X is an arithmetic Veech surface.

3 Characteristic Origamis

3.1 Characteristic Origamis

Definition 3.1 Let **O** be an origami.

a) **O** is called *characteristic* if the subgroup $H(\mathbf{O}) \leqslant \pi_1(E^*, e) \cong F_2$ associated with it is a characteristic subgroup.

b) **O** is called *modular* if its Veech group is the full modular group $\mathrm{SL}_2(\mathbb{Z})$.

From Theorem 2.12, it follows that primitive characteristic origamis are always modular. However, the subgroup of a modular origami need not be stabilized by all inner automorphisms in $\mathrm{Aut}^+(F_2)$, hence it need not be characteristic. We will see that there exist modular origamis that admit no non-trivial translations.

Clearly, every origami **O** is covered by a characteristic origami; it suffices to take the origami associated with $\bigcap_{\varphi \in \mathrm{Aut}^+(F_2)} \varphi(H(\mathbf{O}))$. F. Herrlich's way to construct characteristic origamis [Her06, Proposition 2.1] yields a nicer presentation. We sketch the idea: Let G be a finite group, which can be generated by two elements and let $\mathrm{Epi}(F_2, G)/\mathrm{Aut}(G)$ denote the finite set of epimorphisms $F_2 \to G$ modulo the right action of $\mathrm{Aut}(G)$. If $h_1, \ldots, h_r$ is a system of representatives of $\mathrm{Epi}(F_2, G)/\mathrm{Aut}(G)$, then $\mathrm{Ker}(h) \leqslant F_2$ is a characteristic subgroup, where $h = \prod_{i=1}^r h_i : F_2 \to G^r$.

Trivial examples of characteristic origamis come from certain isogenies $E' \to E$ between elliptic curves: Arrange N^2 squares in one big square and pair opposite horizontal and vertical sides. The resulting origami is characteristic, for the subgroup associated with it is the pullback of the lattice $(\mathbb{Z}/(N))^2$ by the abelianization map $F_2 \to \mathbb{Z}^2$, and $(\mathbb{Z}/(N))^2$ is a characteristic subgroup of $\mathbb{Z}^2$.

3.2 An Infinite Series of Modular Origamis

Let $n \in \mathbb{N}$. Consider the "stairs" origami $\mathbf{St}_n$ [Sch06]. For even n, it is given by

$$\sigma_x = (1\ 2)(3\ 4)\cdots(n-1\ \ n), \quad \sigma_y = (1)(2\ 3)(4\ 5)\cdots(n-2\ \ n-1)(n).$$

and for odd n by

$$\sigma_x = (1\ 2)(3\ 4)\cdots(n-2\ \ n-1)(n), \quad \sigma_y = (1)(2\ 3)(4\ 5)\cdots(n-1\ \ n),$$

We now restrict to odd n. In this case, using the above mentioned idea, Herrlich [Her06] found a characteristic origami $\widetilde{\mathbf{St}}_n$, which covers $\mathbf{St}_n$. We will construct a quotient origami, i. e. an intermediate cover $\widetilde{\mathbf{St}}_n \to \mathbf{N}_n \to E$ with particular properties.

$\widetilde{\mathbf{St}}_n$ corresponds to the subgroup $H(\widetilde{\mathbf{St}}_n)$ of F_2 given in the following way. Let D_n be the dihedral group of order $2n$. D_n has the presentation

$$\langle \sigma, \tau \mid \tau^n, \sigma^2, \sigma\tau\sigma\tau \rangle,$$

and we write elements in the form $\tau^i\sigma^\varepsilon$ for $0 \le i \le n-1$, $\varepsilon \in \{0, 1\}$.

Proposition 3.2 ([Her06, Proposition 4.5])
Let $n \in \mathbb{N}$ be odd, D_n as above and let $h : F_2 = F(\{x, y\}) \to D_n^3$ be the homomorphism given by

$$h(x) = (\sigma, \tau, \sigma), \quad h(y) = (\tau, \sigma, \tau\sigma).$$

Then $\mathrm{Ker}(h) = H(\widetilde{\mathbf{St}}_n)$ *is a characteristic subgroup of F_2, defining an origami $\widetilde{\mathbf{St}}_n$, and $H(\widetilde{\mathbf{St}}_n) \leqslant H(\mathbf{St}_n)$. Moreover, the Galois group of $\widetilde{\mathbf{St}}_n$ is given by the image of h*

$$\mathrm{Gal}(\widetilde{\mathbf{St}}_n / E) = K_n = \big\{(\delta_1, \delta_2, \delta_3) \in D_n^3 \mid e(\delta_1) + e(\delta_2) + e(\delta_3) = 0\big\}$$

where $e : D_n \to \mathbb{Z}/2\mathbb{Z}$ is given by $e(\tau^i\sigma^\varepsilon) = \varepsilon$. It is a primitive origami of degree $4n^3$ in the stratum $\Omega\mathcal{M}_g((n-1)^{4n^2})$, where $g = 2n^2(n-1) + 1$.

Theorem 3.3 *Using the notations of Proposition 3.2, we set*

$$L_n = \langle(\sigma,\sigma,1),(1,\sigma,\sigma)\rangle \cong \mathbb{Z}/(2)\times\mathbb{Z}/(2).$$

Then the origami $\mathbf{N}_n$ *associated with the subgroup* $h^{-1}(L_n)\leqslant F_2$ *has the following properties:*

- *It is a primitive origami given as a gluing of* n^3 *squares. If the squares are labeled by* $(i,j,k)\in(\mathbb{Z}/(n))^3$*, then the right neighbor of* (i,j,k) *is given by*

$$f_h(i,j,k) = (-i, j+1, -k),$$

  and the top neighbor of (i,j,k) *is given by*

$$f_v(i,j,k) = (i+1, -j, 1-k).$$

- *Its genus is* $g=\frac{1}{2}n^2(n-1)+1$ *and* $\mathbf{N}_n$ *lives in the stratum* $\Omega\mathcal{M}_g((n-1)^{n^2})$.
- *Its Veech group* $\Gamma(\mathbf{N}_n)=\mathrm{SL}_2(\mathbb{Z})$ *and its only non-trivial affine biholomorphism is an involution* s *of derivative* $-I$*, given as a permutation of the squares* $\{(i,j,k)\in(\mathbb{Z}/(n))^3\}$ *by*

$$(i,j,k)\mapsto s(i,j,k) = (-i,-j,k).$$

  s *has* n^2+3n *fixed points; the genus of* $\mathbf{N}_n/\langle s\rangle$ *is* $\frac{1}{4}(n(n+1)(n-3))+1$.

PROOF: The short exact sequence

$$1 \longrightarrow \mathrm{Ker}(e\times e\times e) \hookrightarrow D_n^3 \xrightarrow{e\times e\times e} (\mathbb{Z}/(2))^3 \longrightarrow 1$$

splits and induces a presentation of K_n as a semidirect product of $\mathrm{Ker}(e\times e\times e)$ with L_n. A section $S:(\mathbb{Z}/(2))^3\to D_n^3$ is given by $(\varepsilon_1,\varepsilon_2,\varepsilon_3)\mapsto(\sigma^{\varepsilon_1},\sigma^{\varepsilon_2},\sigma^{\varepsilon_3})$, and $L_n = S(\langle(1,1,0),(0,1,1)\rangle)$. Therefore the left cosets of L_n in K_n can be represented by (τ^i,τ^j,τ^k), where $(i,j,k)\in(\mathbb{Z}/(n))^3$. They are in bijection with the left cosets of $h^{-1}(L_n)\leqslant F_2$. Therefore by Proposition 2.10, $\mathbf{N}_n$ is given as the gluing of n^3 squares. The monodromy

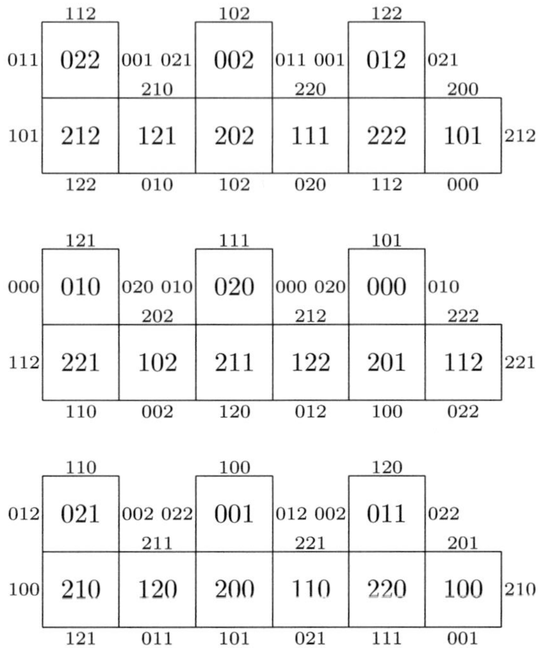

Figure 3.1: *The origami* $\mathbf{N}_3$

of $\mathbf{N}_n$ is given by the action of x and y on the left cosets of $h^{-1}(L_n)$ which is the same as the action of $h(x)$, $h(y)$ on the left cosets of L_n. Let the squares be labeled with the elements of $(\mathbb{Z}/(n))^3$, and identify $(i,j,k) \in (\mathbb{Z}/(n))^3$ with the coset represented by (τ^i, τ^j, τ^k). Since

$$h(x)(\tau^i, \tau^j, \tau^k)L_n = (\sigma\tau^i, \tau\tau^j, \sigma\tau^k)(\sigma, 1, \sigma)L_n = (\tau^{-i}, \tau^{j+1}, \tau^{-k}),$$

and

$$h(y)(\tau^i, \tau^j, \tau^k)L_n = (\tau\tau^i, \sigma\tau^j, \tau\sigma\tau^k)(1, \sigma, \sigma)L_n = (\tau^{i+1}, \tau^{-j}, \tau^{1-k}),$$

the gluings are given as stated above. To compute the genus, we need to know the number of vertices of the squares after identification. They correspond to orbits of the commutator

$$[h(y), h(x)] = h(y)h(x)h(y^{-1})h(x^{-1}) = (\tau^2, \tau^{-2}, \tau^2) =: c$$

Let $m \in \mathbb{Z}$. Since $c^m \cdot (\tau^i, \tau^j, \tau^k)L_n = (\tau^{2m+i}, \tau^{-2m+j}, \tau^{2m+k})L_n$, we see that c^m stabilizes $(\tau^i, \tau^j, \tau^k)L_n$ if and only if

$$2m + i \equiv i \bmod n \ , \quad -2m + j \equiv j \bmod n \ , \quad 2m + k \equiv k \bmod n.$$

Therefore, since n is odd, $m \equiv 0 \bmod n$. So each $\langle c \rangle$-orbit has length n, and there are n^2 orbits; in particular, each vertex of a square is a conical point. The multiplicity of the zero of the holomorphic 1-form at each of the conical points is equal to $n-1$, so $\mathbf{N}_n$ is in the stratum $\Omega\mathcal{M}_g((n-1)^{n^2})$. The formula for the genus g follows from $2g - 2 = (n-1)n^2$.

By Proposition 2.13, the group of translations of $\mathbf{N}_n$ is isomorphic to

$$N_{F_2}(h^{-1}(L_n))/h^{-1}(L_n) \ ,$$

where $N_G(\cdot)$ denotes the normalizer in the group G. Since $\operatorname{Ker}(h)$ is normal in F_2,

$$N_{F_2}(h^{-1}(L_n))/\operatorname{Ker}(h) \cong N_{F_2/\operatorname{Ker}(h)}(h^{-1}(L_n)/\operatorname{Ker}(h)) = N_{K_n}(L_n).$$

So we must show $N_{K_n}(L_n) = L_n$. Let $g \in N_{K_n}(L_n)$ be an element in the normalizer, $g = (\tau^{i_1}\sigma^{\varepsilon_1}, \tau^{i_2}\sigma^{\varepsilon_2}, \tau^{i_3}\sigma^{\varepsilon_3})$. We compute

$$\begin{aligned} g(1,\sigma,\sigma)g^{-1} &= (1, \tau^{i_2}\sigma\tau^{-i_2}, \tau^{i_3}\sigma\tau^{-i_3}) \\ &= (1, \tau^{2i_2}\sigma, \tau^{2i_3}\sigma). \end{aligned}$$

This being in L_n requires $i_2 \equiv i_3 \equiv 0 \bmod n$, since n is odd. In the same way, we see that $i_1 \equiv 0 \bmod n$ by inspecting $g(\sigma,\sigma,1)g^{-1}$. This proves that $\mathbf{N}_n$ has no non-trivial translations.

By Lemma 3.4 below and Theorem 2.12, it suffices to show that any automorphism in $\operatorname{Aut}(F_2/\operatorname{Ker}(h)) = \operatorname{Aut}(K_n)$ leaves the conjugacy class of L_n invariant. But L_n is the 2-Sylow subgroup of K_n. Therefore, the claim follows from the general fact, that in a finite group G, every p-Sylow subgroup is conjugate to every other p-Sylow subgroup, and any element of $\operatorname{Aut}(G)$ maps a p-Sylow subgroup to a p-Sylow subgroup.

We showed above that $\mathbf{N}_n$ has no non-trivial translations. Since $\{\mathrm{id}\} = \operatorname{Trans}(\mathbf{N}_n)$ is an index 2-subgroup of $\operatorname{Aut}(\mathbf{N}_n)$, and since $\Gamma(\mathbf{N}_n) = \mathrm{SL}_2(\mathbb{Z})$, there is precisely one non-trivial affine biholomorphism s, and it is an involution of derivative $-I$. A quick computation shows that the map

$$(i, j, k) \mapsto s(i, j, k) = (-i, -j, k)$$

is a permutation of the square that inverts the edges of the graph $\mathcal{G}(\mathbf{N}_n)$. In fact, we need to check that

$$f_h \circ s = s \circ f_h^{-1} \quad \text{and} \quad f_v \circ s = s \circ f_v^{-1}.$$

Therefore it defines an affine biholomorphism of $\mathbf{N}_n$, which takes the square (i, j, k), rotates it by π, and maps it to $s(i, j, k)$.

We determine the fixed points of s; since s rotates each square by π, they can only be vertices or centers of squares and centers of sides of squares. First, consider the vertices. The lower left vertex of a square corresponds to a $\langle c\rangle$-orbit. s maps the lower left vertex of a square to the upper right vertex of another square. The upper right vertex is the $\langle c\rangle$-orbit of $f_v \circ f_h(i, j, k) = (1 - i, -1 - j, 1 + k)$. Therefore s fixes a vertex, if we can find $m \in \mathbb{Z}$, such that

$$(i, j, k) + m(2, -2, 2) = f_v \circ f_h(s(i, j, k)) = (1 + i, -1 + j, 1 + k).$$

This is equivalent to $m(2, -2, 2) = (1, -1, 1)$, which has a solution, since n is odd. Since this holds for any $(i, j, k) \in (\mathbb{Z}/(n))^3$, each of the n^2 vertices is fixed. The remaining fixed points are considerably easier: A center is fixed, if and only if $s(i, j, k) = (i, j, k)$, which leads to $i = j = 0$ and $k \in \mathbb{Z}/(n)$ arbitrary. A center of a lower side of a square is fixed, if and only if s interchanges (i, j, k) and $f_v(i, j, k)$, which means $s(i, j, k) = f_v^{-1}(i, j, k)$, so $i = k = 2^{-1}$ and $j \in \mathbb{Z}/(n)$ arbitrary. Similarly, a center of a left side of a square is fixed if and only if $s(i, j, k) = f_h^{-1}(i, j, k)$, so $j = 2^{-1}$, $k = 0$ and $i \in \mathbb{Z}/(n)$ is arbitrary. In total, we obtain $n^2 + 3n$ fixed points. Plugging this into the Hurwitz formula yields

$$2g(\mathbf{N}_n) - 2 = n^2(n - 1) = 2(2g(\mathbf{N}_n/\langle s\rangle) - 2) + n^2 + 3n,$$

which is equivalent to $g(\mathbf{N}_n/\langle s\rangle) = \frac{1}{4}(n(n + 1)(n - 3)) + 1$. ■

The following lemma completes the proof of Theorem 3.3. We use the notation $[\cdot]$ for the conjugacy class (of a subgroup) in F_2; moreover, $x^g := gxg^{-1}$ for $x, g \in F_2$.

Lemma 3.4 *Let* $\Phi \in \mathrm{Out}^+(F_2)$, f *a lift of* Φ *to* $\mathrm{Aut}^+(F_2)$ *and* $H \leqslant K \leqslant F_2$ *subgroups, such that* $f(H) = H$. *Assume further that* H *is normal in* F_2. *Then* $\Phi([K]) = [K]$ *if and only if* $\pi(f)(K/H)$ *is conjugate to* K/H *in* F_2/H, *where*

$$\pi : \mathrm{Stab}_{F_2}(H) \to \mathrm{Aut}(F_2/H) \ , \quad f \mapsto (xH \mapsto f(x)H).$$

PROOF: If $\Phi([K]) = [K]$, then there exists $g \in F_2$ such that $f(K) = K^g$. Then $\pi(f)(K/H) = f(K)/H = K^g/H = (K/H)^{gH}$, proving the "only if"-part. Conversely, let $\pi(f)(K/H) = (K/H)^{gH}$ for some $g \in F_2$. We show that $f(K) = K^g$. Let $x \in f(K)$. Then

$$xH \in f(K)/H = \pi(f)(K/H) = (K/H)^{gH} = K^g/H,$$

so there exists $k \in K$, $h \in H$, such that $x = k^g h = (kh^{g^{-1}})^g \in K^g$. If in turn we start with $x \in K^g$, then

$$xH \in K^g/H = (K/H)^{gH} = \pi(f)(K/H) = f(K)/H,$$

so there exists $y \in K$, $h \in H$ with $x = f(y)h = f(yf^{-1}(h))$, which implies $x \in f(K)$, since $yf^{-1}(h) \in K$. ■

3.3 Schmoll's Modular Fibers

There is another source of modular origamis, coming from Schmoll's modular fibers. We shortly describe how they arise. Consider the set F_d of tuples $((X, \omega), \pi, z_1, z_2)$, where

- $\pi : (X, \omega) \to (E, \mathrm{d}z)$ is a translation cover of the square torus $E = \mathbb{C}/(\mathbb{Z} \oplus i\,\mathbb{Z})$ of degree d,
- π is ramified at 2 points $z_1 \neq z_2 \in X$ with multiplicity 2, and $\pi(z_1) = \overline{0}$,
- π is primitive, i. e. $\mathrm{Per}(\omega) = \mathbb{Z} \oplus i\,\mathbb{Z}$.

The image of F_d in $\Omega\mathcal{M}_{2,2}(1, 1)$ can be endowed with a translation structure via the map

$$\Phi_d : F_d \to E \ , \quad ((X, \omega), z_1, z_2) \mapsto \int_{z_1}^{z_2} \omega \bmod \mathrm{Per}(\omega).$$

Schmoll [Sch05b] gave a formula for the number of zeros of $\Phi_d^*(\mathrm{d}z)$ (each one is of order 2), and he showed that $\Gamma(F_d, \Phi_d^*(\mathrm{d}z)) = \mathrm{SL}_2(\mathbb{Z})$. The monodromy of F_d can in theory be derived from [EMS03], but to give a closed formula is, in the author's opinion, very hard. Therefore, we do not know if F_d possesses any non-trivial translations, except for the case $d = 3$, where there are none. Also, it would be interesting to classify the orbits of the $\mathrm{SL}_2(\mathbb{Z})$-action on F_d, for this would yield a classification of the Teichmüller curves in $\Omega\mathcal{M}_2(1,1)$, which are generated by origamis (completing the classification of genus 2). Note that these curves are Hurwitz spaces and have also been described from an algebro-geometric point of view by Kani [Kan03], [Kan06].

4 Homology and Cohomology

We summarize some results for the homology and cohomology of a compact manifold that we will use in the subsequent chapters. A general reference for this part is [Hat02] and [Voi02].

4.1 Singular Homology and Cohomology

For a topological space X, let $H_k^{\mathrm{sing}}(X, \mathbb{Z})$ denote the k-th singular homology group. For the k-th singular cohomology group with coefficients in the abelian group G, we write $H^k_{\mathrm{sing}}(X, G)$.

Remark 4.1

a) By the universal coefficient theorem, we have

$$H^1_{\mathrm{sing}}(X, G) \cong \mathrm{Hom}_{\mathbb{Z}}(H_1^{\mathrm{sing}}(X, \mathbb{Z}), G),$$

and the same also holds for $k \geq 1$ if for example $H_{k-1}^{\mathrm{sing}}(X, \mathbb{Z})$ is free (see [Hat02, Thm 3.2]).

b) Recall that for a path-connected space X, the group $H_1^{\mathrm{sing}}(X, \mathbb{Z})$ is isomorphic to the abelianization of the fundamental group $\pi_1(X)$.

On the homology of a surface, we dispose of a symplectic pairing. Before introducing it, let us recall some generalities on symplectic vector spaces. Let V be a $2g$-dimensional vector space over a field $\mathbb{K}$, not of characteristic 2, or a finitely generated, free $\mathbb{Z}$-module of rank $2g$. Let ω be a symplectic form on V, i. e. a non-degenerate bilinear, alternating pairing with values in the coefficient ring. Then ω induces an identification of V with its dual V^* by

$$\Phi : V \to V^* , \quad a \mapsto \omega(\cdot, a).$$

We obtain a symplectic form ω^* on V^* by setting

$$\omega^*(\lambda,\mu) = \omega(\Phi^{-1}(\lambda), \Phi^{-1}(\mu)),$$

sometimes called Poisson bracket. A linear map $f: V \to W$ such that

$$\omega_W(f(a), f(b)) = \omega_V(a,b)$$

is called *symplectic map*. Note that symplectic maps are necessarily injective. The symplectic form induces an involutive map on subspaces. For $U \leqslant V$ a subspace (or sub-module), let

$$U^\perp = \{v \in V \mid \omega(u,v) = 0 \ \text{ for all } \ u \in U\}.$$

Example 4.2 Let X be a closed surface of genus $g \geq 1$. Then

$$H_0^{\text{sing}}(X,\mathbb{Z}) \cong \mathbb{Z}\ , \quad H_1^{\text{sing}}(X,\mathbb{Z}) \cong \mathbb{Z}^{2g}\ , \quad H_2^{\text{sing}}(X,\mathbb{Z}) \cong \mathbb{Z}\,.$$

Recall that we have a symplectic pairing on $H_1^{\text{sing}}(X,\mathbb{Z})$, the algebraic intersection number i, which assigns to a pair a, $b \in H_1^{\text{sing}}(X,\mathbb{Z})$, represented by closed curves α and β, the number of positive intersections of α and β minus the number of negative intersections. Here, we assume that we have chosen a fundamental class $[X] \in H_2^{\text{sing}}(X,\mathbb{Z})$, thus an orientation on X, and that α and β intersect transversally.

By dualizing, we obtain a symplectic pairing i^* on cohomology. One can show that this pairing coincides with the cup product pairing on cohomology

$$Q: H^1_{\text{sing}}(X,\mathbb{Z}) \times H^1_{\text{sing}}(X,\mathbb{Z}) \to \mathbb{Z}\ , \quad (\varphi,\psi) \mapsto (\varphi \cup \psi)([X]).$$

Poincaré duality tells us that Q is non-degenerate, while i is non-degenerate by surface surgery arguments, namely representing the closed surface X as being glued from a regular $4g$-gon.

In particular, let $f: X \to Y$ be a (ramified) covering map of surfaces. It induces $f_*: H_1^{\text{sing}}(X,\mathbb{Z}) \to H_1^{\text{sing}}(Y,\mathbb{Z})$ and $f^*: H^1_{\text{sing}}(Y,\mathbb{Z}) \to H^1_{\text{sing}}(X,\mathbb{Z})$. The image of f_* is of finite index in $H_1^{\text{sing}}(Y,\mathbb{Z})$. Moreover, f^* is a symplectic map for the symplectic forms i_X^* and $\deg(f) i_Y^*$, i. e.

$$i_X^*(f^*\lambda, f^*\mu) = \deg(f)\cdot i_Y^*(\lambda,\mu)$$

for all λ, $\mu \in H^1_{\text{sing}}(Y,\mathbb{Z})$.

On a topological space X, we also dispose of the sheaf cohomology. We denote by $H^k(X, \mathcal{F})$ the cohomology groups with values in the sheaf $\mathcal{F}$ (see [Voi02, Chap. 4]).

Proposition 4.3 ([Voi02, Thm 4.47])
On a locally contractible topological space X we have a canonical isomorphism

$$H^k_{\mathrm{sing}}(X, R) \simeq H^k(X, R_X) = H^k(X, R)$$

for any commutative Ring R.

Remark 4.4 One more remark concerning the change of coefficients (see [Voi02, p. 157]). If X is a compact manifold and if R is a field of characteristic 0, then there is a canonical isomorphism

$$H^1(X, \mathbb{Z}) \otimes_{\mathbb{Z}} R \cong H^1(X, R).$$

4.2 De Rham Cohomology

In case X is a differentiable (i. e. C^∞) manifold, there is another description of $H^k(X, \mathbb{R})$, respectively $H^k(X, \mathbb{C})$ in terms of classes of differential forms.

Let $\mathbb{K} = \mathbb{R}$ or $\mathbb{C}$ and let $\mathcal{A} = \mathcal{A}_X$ denote the sheaf of C^∞-differential forms on X with values in $\mathbb{K}$, i. e. the sheaf of C^∞-sections of the real or complexified cotangent bundle. Let $\mathcal{A}^k = \bigwedge^k \mathcal{A}$ ($k \in \mathbb{N}_0$) denote the corresponding k-forms, and let d be the differential. The cochain complex

$$0 \longrightarrow \mathcal{A}^0 \xrightarrow{\ d\ } \mathcal{A}^1 \xrightarrow{\ d\ } \dots \xrightarrow{\ d\ } \mathcal{A}^n \to 0$$

(where $n = \dim X$) is a resolution of the constant sheaf $\mathbb{K}$, hence the associated cochain complex that we obtain by taking global sections computes the sheaf cohomology of $\mathbb{K}$. This is the de Rham Theorem.

Proposition 4.5 ([Voi02, Thm. 4.1])
For a differentiable manifold X

$$\begin{aligned} H^k_{dR}(X) &:= \mathrm{Ker}(d : \mathcal{A}^k(X) \to \mathcal{A}^{k+1}(X)) / \mathrm{Im}(d : \mathcal{A}^{k-1}(X) \to \mathcal{A}^k(X)) \\ &\cong H^k(X, \mathbb{K}). \end{aligned}$$

Remark 4.6 ([Voi02, Thm. 5.29])
In particular, if X is a compact, oriented, connected n-dimensional differentiable manifold, then the pairing given by cup product and Poincaré duality

$$H^k(X,\mathbb{K}) \times H^{n-k}(X,\mathbb{K}) \to \mathbb{K}$$

can be rewritten using the above identification of $H^k(X,\mathbb{K})$ with $H^k_{dR}(X)$ as

$$(\alpha,\beta) \mapsto \int_X \alpha \wedge \beta,$$

where we choose representatives α and β of classes in $H^k_{dR}(X)$, respectively $H^{n-k}_{dR}(X)$. Note that in the case $n=2$, $k=1$ we obtain the pairing from Example 4.2.

4.3 Hodge Decomposition

We describe the Hodge decomposition of the cohomology groups $H^k(X,\mathbb{C})$ of a compact Kähler manifold. This is a realization of the abstract concept of polarised Hodge structures, which we will encouter later. In this section, let again $\mathbb{K} = \mathbb{R}$ or $\mathbb{C}$.

Harmonic Forms

If X is a compact, oriented, connected n-dimensional Riemannian manifold, we obtain an even better description of $H^k(X,\mathbb{K})$. With the help of the Riemannian metric g on X, one defines the Laplacian

$$\Delta : \mathcal{A}^k(X) \to \mathcal{A}^k(X).$$

The subspace $\operatorname{Ker}(\Delta) \subset \mathcal{A}^k(X)$ is the space $\mathcal{H}^k = \mathcal{H}^k_{\mathbb{K}}$ of harmonic k-forms (with values in $\mathbb{K}$).

Proposition 4.7 ([Voi02, Thm. 5.2])
The linear map

$$\mathcal{H}^k_{\mathbb{K}} \to H^k(X,\mathbb{K}) \ , \quad \omega \mapsto [\omega] \ ,$$

sending ω to its de Rham cohomology class, is an isomorphism.

Kähler Manifolds

Recall that a Kähler manifold is a complex manifold X, whose tangent bundle carries a (positive definite) hermitian metric $h = g - i\omega$, i. e. g is a Riemannian metric, and ω is a closed real 2-form of type $(1,1)$ – the Kähler form.

Example 4.8 Let X be a Riemann surface. Then X is a 2-dimensional real manifold, so every 2-form is closed. Therefore X is a Kähler manifold with any choice of a hermitian metric on X.

In the following, let X be a compact Kähler manifold. The complex structure on the tangent bundle of X induces a decomposition $\mathcal{A} = \mathcal{A}^{1,0} \oplus \mathcal{A}^{0,1}$, which in turn allows us to decompose $\mathcal{A}^k$ into

$$\mathcal{A}^k = \bigoplus_{p+q=k} \mathcal{A}^{p,q}\,.$$

From the Kähler identities, one can deduce that this decomposition also descends to the harmonic k-forms, so that

$$\mathcal{H}^k_{\mathbb{C}} = \bigoplus_{p+q=k} \mathcal{H}^{p,q}\,.$$

Remark 4.9 ([Voi02, Cor. 6.10, Prop. 6.11])

a) Via the identification in Proposition 4.7, we obtain a decomposition of $H^k(X,\mathbb{C})$, the Hodge decomposition

$$H^k(X,\mathbb{C}) = \bigoplus_{p+q=k} H^{p,q}.$$

b) This decomposition is independent of the choice of a particular Kähler metric on X.

c) We have $\overline{H^{p,q}} = H^{q,p}$, where complex conjugation acts on the second factor of

$$H^k(X,\mathbb{R}) \otimes_{\mathbb{R}} \mathbb{C} \cong H^k(X,\mathbb{C})$$

in the usual way.

Let Ω_X^1 denote the sheaf of holomorphic 1-forms on X, and Ω_X^p the sheaf of holomorphic p-forms, i. e. the p-th exterior power of Ω_X^1.

Proposition 4.10 ([Voi02, Remark 8.29])
Let X be a compact Kähler manifold. We have an isomorphism of complex vector spaces

$$H^{p,q} \cong H^q(X, \Omega_X^p),$$

which does not depend on the chosen Kähler metric.

In particular, $H^{1,0} \cong \Omega_X^1(X)$.

4.4 Riemann Surfaces

Let X be a compact Riemann surface. To bring us in the above setup, choose a Kähler metric on X. Note however that the choice does not matter as everything can also be formulated without recourse to the metric (see [For81, Chap. 19]).

In this case, we see from Remarks 4.9 and 4.10, that

$$H^1(X, \mathbb{C}) \cong \Omega_X^1(X) \oplus \overline{\Omega_X^1(X)}.$$

Proposition 4.11 ([For81, Thm. 19.4])
The map

$$\Omega_X^1(X) \to \mathcal{H}_{\mathbb{R}}^1 , \quad \omega \mapsto \mathrm{Re}\,(\omega)$$

is an isomorphism of real vector spaces (as is the map, which sends a holomorphic 1-form to its imaginary part).

Recall from Remark 4.6 that we dispose of a non-degenerate, alternating pairing on $H^1(X, \mathbb{C})$ given by

$$(\alpha, \beta) \mapsto Q(\alpha, \beta) = \int_X \alpha \wedge \beta.$$

We can modify this pairing to obtain a positive definite hermitian form on $H^1(X, \mathbb{C})$. Consider the Hodge $*$-operator. In this simple setup, it is the $\mathbb{C}$-linear automorphism

$$* : \mathcal{A}^1(X) \to \mathcal{A}^1(X) , \quad \omega = \omega^{1,0} + \omega^{0,1} \mapsto i(\overline{\omega^{1,0}} - \overline{\omega^{0,1}}),$$

where we decompose ω into $\omega^{1,0} \in \mathcal{A}^{1,0}(X)$ and $\omega^{0,1} \in \mathcal{A}^{0,1}(X)$. It descends to harmonic 1-forms, thus also to $H^1(X,\mathbb{C})$ by Proposition 4.7.

Remark 4.12 Set

$$H : H^1(X,\mathbb{C}) \times H^1(X,\mathbb{C}) \to \mathbb{C}\ , \quad (\alpha,\beta) \mapsto \int_X \alpha \wedge *\beta.$$

Then H is a positive definite hermitian form, for which the Hodge decomposition is orthogonal. Its restriction to $H^{1,0}$ is given by

$$(\alpha,\beta) \mapsto i \int_X \alpha \wedge \overline{\beta}.$$

Via the identification in Proposition 4.11, we obtain a norm on $H^1(X,\mathbb{R})$, the *Hodge norm*, by setting

$$\|\tilde{\alpha}\| = \sqrt{H(\alpha,\alpha)}\ ,$$

where α is the preimage of $\tilde{\alpha} \in H^1(X,\mathbb{R})$ under the map Re.

5 Families of Curves

Following A. Grothendieck [Gro62a], we set up the notion of a family of curves. We will work in the analytic category (i. e. the category of complex spaces), so if we speak of algebraic objects, such as stable curves, we secretly apply the functor from schemes over $\mathbb{C}$ to complex spaces to them.

Definition 5.1 Let $\mathcal{X}$, B be complex spaces, and let $\phi : \mathcal{X} \to B$ be a morphism.

a) ϕ is called *family of curves of genus* g, if ϕ is a proper, flat morphism with smooth fibers, which are compact Riemann surfaces of genus g.

b) ϕ is called *family of stable curves of genus* g, if ϕ is proper and flat, and its fibers are stable curves of arithmetic genus g.

Remark 5.2 a) If the base B is smooth, i. e. a complex manifold, then a family $\phi : \mathcal{X} \to B$ of curves of genus g is a proper submersion between complex manifolds.

b) A family $\phi : \mathcal{X} \to B$ of curves of genus g is locally topologically trivial, i. e. every point $b \in B$ has a neighborhood $U \subset B$, such that there exists a homeomorphism $h : \phi^{-1}(U) \to U \times \mathcal{X}_b$ over U. Here $\mathcal{X}_b = \phi^{-1}(b)$.

c) If the base is smooth, then we can also find a C^∞-trivialization in b).

Part a) follows from [Gro62b, Théorèm 3.1]. Part b) is proved in [Gro62c, Proposition 1.8]. Part c) is widely known, see e. g. [Voi02, Proposition 9.5].

The next remark collects some facts about moduli spaces. References on this topic are e. g. [DM69], [HM98].

Remark 5.3 a) We can consider the moduli space $\mathcal{M}_g$, respectively its Deligne-Mumford compactification $\overline{\mathcal{M}}_g$ as the stack over the category of complex spaces, whose fiber over B is the gruppoid of families of curves of genus g, respectively families of stable curves of genus g.

b) $\mathcal{M}_g$ and $\overline{\mathcal{M}}_g$ are not representable. However, if we rigidify the moduli problem, then we obtain finite covers that are representable by complex manifolds.

c) In particular, consider families of curves of genus g with a level-ℓ structure, by which we understand the choice of an isomorphism from $(\mathbb{Z}/(\ell))^{2g}$ to the ℓ-torsion points of the Jacobian. If $\ell \geq 3$, then there is a fine moduli space representing this moduli problem, namely

$$\mathcal{T}_g \,/\, \Gamma_g^{[\ell]},$$

where $\Gamma_g^{[\ell]}$ is the kernel of the action of the mapping class group Γ_g on $H_1(S, \mathbb{Z}/(\ell))$.

d) For $n \in \mathbb{N}$, E. Looijenga [Loo94] constructed a finite cover $\overline{\mathcal{M}}_g[(\binom{n}{2})]$ of $\overline{\mathcal{M}}_g$, parametrizing curves with a Prym level structure. He showed that for $n \geq 6$, $\overline{\mathcal{M}}_g[(\binom{n}{2})]$ is smooth, i. e. a fine moduli space. $\overline{\mathcal{M}}_g[(\binom{n}{2})]$ is the normalization in $\overline{\mathcal{M}}_g$ of the quotient $\mathcal{T}_g \,/\, \Gamma_{g,(\binom{n}{2})}$, where $\Gamma_{g,(\binom{n}{2})}$ is a certain finite index subgroup of Γ_g.

Remark 5.4 The usefulness of having fine moduli spaces at hand, consists in the existence of a universal curve. Let (X, ω) be a Veech surface of genus g, and let

$$\mathbb{H} \,/\, \mathrm{Aff}(X, \omega) \to \mathcal{M}_g$$

be the associated Teichmüller curve. By intersection of $\mathrm{Aff}(X, \omega)$ inside Γ_g with the above mentioned subgroups, we can find a subgroup $\Gamma \leqslant \mathrm{Aff}(X, \omega)$ fulfilling the condition

> **Condition (∗).** $\Gamma \leqslant \mathrm{Aff}(X, \omega)$ is a finite-index subgroup, such that $j : \mathcal{C} = \mathbb{H}/\Gamma \to \mathcal{M}_g$ factors over a fine moduli space. Let $\overline{\mathcal{C}}$ be the completion of $\mathcal{C}$ (i. e. the normalization of $\mathcal{C}$ inside $\overline{\mathcal{M}}_g$). The pullback of the universal curve over a suitable cover of $\overline{\mathcal{M}}_g$ to $\overline{\mathcal{C}}$ is a family $\overline{\phi} : \overline{\mathcal{X}} \to \overline{\mathcal{C}}$ of stable curves of genus g. Its restriction to $\mathcal{C}$ is a family $\phi : \mathcal{X} \to \mathcal{C}$ of curves of genus g.

In particular, we remark that Γ is torsion-free, thus it is the fundamental group of $\mathcal{C}$. Moreover, the local monodromy of Γ about the cusps is unipotent, and not just quasi-unipotent (see Section 7.3).

Definition 5.5 The family $\phi : \mathcal{X} \to \mathcal{C}$ of curves of genus g of Condition $(*)$ is called *family over the Teichmüller curve* associated with (X, ω).

For the next proposition, we need to recall the construction of the universal curve over Teichmüller space. We present the Bers fiber space approach. Let X be a compact Riemann surface of negative Euler characteristic, and let $M(X)$ be the space of Beltrami forms on X (see e. g. [Hub06]). Fix a universal cover $\mathbb{H} \to X$ and let π be the group of deck transformations. There is a natural identification of $M(X)$ with $M^{\pi}(\mathbb{H})$, the space of π-invariant Beltrami forms on $\mathbb{H}$. For $\mu \in M(X)$, let $\hat{\mu}$ be the extension to $\mathbb{C}$ by 0 of the lift of μ to $\mathbb{H}$, and let $f^{\hat{\mu}} : \mathbb{C} \to \mathbb{C}$ be the solution of the Beltrami equation for $\hat{\mu}$, normalized to fix 0, 1, and ∞. Consider the map

$$\Psi : M(X) \times \mathbb{H} \to \mathcal{T}(X) \times \mathbb{C}, \quad (\mu, z) \mapsto (\Phi_X(\mu), f^{\hat{\mu}}(z)) ,$$

where $\Phi_X : M(X) \to \mathcal{T}(X)$ is the projection [Hub06, Sect. 6.4]. The group π acts holomorphically on $\mathrm{Im}(\Psi)$ by the rule $(\Phi_X(\mu), z) \mapsto (\Phi_X(\mu), f^{\hat{\mu}} \circ c \circ (f^{\hat{\mu}})^{-1}(z))$ for $c \in \pi$. The universal family over $\mathcal{T}(X)$ is given by $\mathrm{Im}(\Psi)/\pi \to \mathcal{T}(X)$.

Proposition 5.6 *Let $p : (X, \omega) \to (Y, \nu)$ be a translation cover between translation surfaces. Let B be the branch locus of p, and let $Y^* = Y \setminus B$ and $X^* = X \setminus p^{-1}(B)$. Suppose that Y^* has negative Euler characteristic. Let $\jmath_{X^*} : \mathbb{H} \to \mathcal{T}(X, p^{-1}(B))$ and $\jmath_{Y^*} : \mathbb{H} \to \mathcal{T}(Y, B)$ be the corresponding Teichmüller disks, and let $\phi_{X^*} : \tilde{\mathcal{X}}^* \to \mathbb{H}$, respectively $\phi_{Y^*} : \tilde{\mathcal{Y}}^* \to \mathbb{H}$ be the pullback of the universal family over Teichmüller space. Then p induces a holomorphic map $F : \tilde{\mathcal{X}}^* \to \tilde{\mathcal{Y}}^*$ over $\mathbb{H}$.*

Proof: Fix a universal cover $\mathbb{H} \to Y^*$, which factors over $p : X^* \to Y^*$. Let $\pi_X \leqslant \pi_Y$ be the deck transformation groups of $\mathbb{H} \to X^*$ and $\mathbb{H} \to Y^*$ respectively. To ease notation, we will use $\mathbb{D}$ in place of $\mathbb{H}$. Also note that Beltrami forms are measurable functions, and do not care about the nullsets B and $p^{-1}(B)$. The Teichmüller disk of Y is given as the projection of the image of

$$\mathbb{D} \to M(Y)\ , \quad t \mapsto \mu_t = t\frac{\overline{q}}{|q|} = t\frac{\overline{\nu}}{\nu},$$

where $q = \nu^2$. Similarly, the Teichmüller disk of X is the projection of the image of

$$\mathbb{D} \to M(X)\ , \quad t \mapsto t\frac{\overline{\omega}}{\omega} = p^*\big(t\frac{\overline{\nu}}{\nu}\big).$$

In particular, for any $t \in \mathbb{D}$, pulling back the differentials $t\frac{\overline{\nu}}{\nu}$ and $t\frac{\overline{\omega}}{\omega}$ to $\mathbb{H}$ yields the same Beltrami form in $M(\mathbb{H})$. We now consider the restriction

$$\Psi : \mathbb{D} \times \mathbb{H} \to \mathbb{D} \times \mathbb{C}, \quad (t, z) \mapsto (t, f^{\hat{\mu}_t}(z)).$$

It produces simultaneously the families $\tilde{\mathfrak{X}}^*$ and $\tilde{\mathfrak{Y}}^*$; the first by factoring out π_X, the latter by factoring out π_Y in the image of Ψ. The inclusion $\pi_X \leqslant \pi_Y$ induces a holomorphic map $F : \tilde{\mathfrak{X}}^* \to \tilde{\mathfrak{Y}}^*$ over $\mathbb{D} \cong \mathbb{H}$, such that the map between the fibers over 0 is given by $p : X^* \to Y^*$. ■

6 Local Systems, Monodromy Representations and Vector Bundles with a Flat Connection

We present a triptych of equivalent categories that one can define on a complex manifold B: local systems, π_1-representations and vector bundles with a flat connection. Being able to pass freely from one description to another will be crucial in the arguments of the subsequent chapters. We follow the presentation of Deligne [Del70, Chap. 1] (without mentioning the point of view of differential equations which is also implicit).

Throughout this chapter, let B be a complex manifold.

6.1 Local Systems

Consider a family of curves $\phi : \mathfrak{X} \to B$. The cohomology groups $H^1(\mathfrak{X}_b, \mathbb{Z})$ of the fibers $\mathfrak{X}_b$ ($b \in B$) can be glued together in a sense that is to be made precise in the following section. They form a local system.

Definition 6.1 Let X be a locally connected topological space and let M be a module over a ring R.

a) A *local system* $\mathbb{V}$ of stalk M on X, is a sheaf of R-modules on X, which is locally isomorphic to the constant sheaf M_X.

b) By an R-local system we shall usually understand a local system with stalk a finitely generated R-module M.

c) The category of R-local systems is the full subcategory of the category of sheaves of R-modules on X.

Local systems behave like vector bundles in the following sense.

Remark 6.2 We present an alternative way to view and construct local systems. Let $\{U_i\}$ be an open covering of X. On $U_i \cap U_j$, assume that we are given a cocycle $g_{ij} : U_i \cap U_j \to \mathrm{Aut}(M)$, i. e. a locally constant map, such that on $U_i \cap U_j \cap U_k$, we have

$$g_{jk} \circ g_{ij} = g_{ik}.$$

Then we can glue together the patches $U_i \times M$ by identifying $(u, v) \in U_i \times M$ with $(u, g_{ij}(u)(v)) \in U_j \times M$ if $u \in U_i \cap U_j$, to obtain a topological space T with a projection $\pi : T \to X$. The locally constant local sections of π are the local sections of the local system $\mathbb{M}$.

There is a canonical local system associated with families of curves.

Lemma 6.3 *Let $\phi : \mathcal{X} \to B$ be a family of curves of genus g, let A be a ring, and let $k \geq 0$. Let $R^k\phi_*$ be the k-th right derived functor of the functor ϕ_* from the category of sheaves of abelian groups on $\mathcal{X}$ to the category of sheaves of abelian groups on B.*

a) The sheaf $R^k\phi_*(A_\mathcal{X})$ is the sheaf on B associated to the presheaf

$$U \mapsto H^k(\phi^{-1}(U), A).$$

b) $R^k\phi_*(A_\mathcal{X})$ is an A-local system.

PROOF: a) see [Har04, Proposition III.8.1]

b) Let U be a contractible neighborhood of a point $b \in B$, where ϕ has a C^∞-trivialization $h : U \times \mathcal{X}_b \to \phi^{-1}(U)$. Then by Proposition 4.3,

$$\begin{aligned} H^k(\phi^{-1}(U), A) &\cong H^k_{\mathrm{sing}}(\phi^{-1}(U), A) \cong H^k_{\mathrm{sing}}(U \times \mathcal{X}_b, A) \\ &\cong H^k_{\mathrm{sing}}(\mathcal{X}_b, A) \cong H^k(\mathcal{X}_b, A). \end{aligned}$$

By inspecting which sections are in the sheafification of $H^k(\phi^{-1}(\cdot), A)$, one finds that $R^k\phi_*(A_\mathcal{X})$, restricted to U, is isomorphic to the constant sheaf of stalk $H^k(\mathcal{X}_b, A)$. ∎

6.2 Monodromy Representation

First, let us state that a local system on a sufficiently well-behaved space is the same as a representation of the fundamental group.

Proposition 6.4 *Let R be a ring, and let X be a path-connected, locally simply connected topological space with base point x. Then there is an equivalence between the category of R-local systems on X and the category of $\pi_1(X,x)$-left modules, given by the functor*

$$\mathbb{V} \mapsto \mathbb{V}_x\,.$$

PROOF: We sketch the essential steps (see also [Voi03, Sect.3.1.1]). First, given a path $c : [0,1] \to X$, starting at $x = c(0)$, there is a unique way of continuing a germ $v \in \mathbb{V}_x$ along γ to an element $v' \in \mathbb{V}_{c(1)}$ (since every germ in $\mathbb{V}_x$ produces a unique section $\mathbb{V}(U)$ for some neighborhood U of x). This continuation process only depends on the homotopy class. Thus, it allows us to define a representation $\pi_1(X,x) \to \mathrm{Aut}(\mathbb{V}_x)$.

To construct the inverse functor, we start with a representation

$$\rho : \pi_1(X,x) \to \mathrm{Aut}(V)$$

(with V an R-module) and consider the constant sheaf $V_{\tilde{X}}$ on the universal cover $u : \tilde{X} \to X$. We define an R-local system $\mathbb{V}$ on X by taking on an open set $U \subset X$ the sections $f : u^{-1}(U) \to V$ of $V_{\tilde{X}}$ that satisfy

$$f(\gamma \cdot x) = \rho(\gamma) f(x)$$

for all $\gamma \in \pi_1(X,x)$, $x \in u^{-1}(U)$. Then $\mathbb{V}$ is isomorphic to the constant sheaf V_U on a sufficiently small neighborhood $U \subset X$ (where $u^{-1}(U) = \bigcup_{\gamma \in \pi_1(X,x)} V_\gamma$ with disjoint open sets V_γ homeomorphic to U), so it is a local system. We leave the rest to the reader. ∎

Definition 6.5 Let $\phi : \mathfrak{X} \to B$ be a family of curves, let $b \in B$, and A be a ring. The $\pi_1(B,b)$-left module $\mathbb{V}_b$ associated with the local system $\mathbb{V} = R^1\phi_*(A_{\mathfrak{X}})$ is called the *monodromy representation* of the family ϕ (with values in A).

Example 6.6 Let (X,ω) be a Veech surface, and let $f \in \mathrm{Aff}(X,\omega)$. Then f_* acts on $H_1(X,\mathbb{Z})$ from the left, preserving the intersection form i. By dualizing, we obtain the action

$$\rho : \mathrm{Aff}(X,\omega) \to \mathrm{Aut}(H^1(X,\mathbb{Z}))\ , \quad f \mapsto (f^*)^{-1}$$

which in fact lands in $\mathrm{Sp}(H^1(X,\mathbb{Z})) \leqslant \mathrm{Aut}(H^1(X,\mathbb{Z}))$, since $(f^{-1})^*$ preserves the intersection form i^* of Example 4.2.

Let $j : \mathcal{C} = \mathbb{H} / \mathrm{Aff}(X,\omega) \to \mathcal{M}_g$ be the Teichmüller curve to (X,ω). Let Γ be a subgroup of finite index of $\mathrm{Aff}(X,\omega)$ satisfying Condition $(*)$, and consider the finite cover $\mathcal{C}' = \mathbb{H}/\Gamma$ of $\mathcal{C}$. As Γ acts freely on $\mathbb{H}$, it is the fundamental group of $\mathcal{C}'$. Let $c' \in \mathcal{C}'$ be a point that is mapped to $[X] \in \mathcal{M}_g$. Then the restriction of ρ to Γ

$$\rho : \Gamma \to \mathrm{Sp}(H^1(X,\mathbb{Z}))$$

is a monodromy representation (see [Bau09, Lemma 2.4.3]).

Note that if we tensor ρ by a field $\mathbb{K}$ of characteristic 0, we obtain a representation $\rho \otimes_{\mathbb{Z}} \mathbb{K} : \pi_1(\mathcal{C}',c') \to \mathrm{Sp}(H^1(X,\mathbb{K}))$ by Remark 4.4.

By abuse of language, we will also call $\rho : \mathrm{Aff}(X,\omega) \to \mathrm{Sp}(H^1(X,\mathbb{Z}))$ a monodromy representation, even if $\mathrm{Aff}(X,\omega)$ is only the orbifold fundamental group of the Teichmüller curve.

Another cool fact about the representation of $\mathrm{Aff}(X,\omega)$ above is the following.

Remark 6.7 The action $\rho : \mathrm{Aff}(X,\omega) \to \mathrm{Sp}(H^1(X,\mathbb{Z}))$ is faithful (see e. g. [Bau09, Lemma 2.3.17]).

Remark 6.8 Let $\mathbf{O}$ be an origami. Consider a surface $A \cdot \mathbf{O} \neq \mathbf{O}$ in the $\mathrm{SL}_2(\mathbb{Z})$-orbit of $\mathbf{O}$ (which is again an origami). Then the difference between the monodromy representation of $\mathbf{O}$ and the one of $A{\cdot}\mathbf{O}$ is roughly a change of the base point of the fundamental group (with the slight imprecision that we are dealing only with orbifold fundamental groups). More precisely, let $\varphi_A : \mathbf{O} \to A \cdot \mathbf{O}$ be the affine shear from Section 2.2.

Let $\rho : \mathrm{Aff}(\mathbf{O}) \to \mathrm{Sp}(H^1(\mathbf{O}, \mathbb{Z}))$ and $A \cdot \rho : \mathrm{Aff}(A \cdot \mathbf{O}) \to \mathrm{Sp}(H^1(A \cdot \mathbf{O}, \mathbb{Z}))$ be the monodromy representations. Then the diagram

$$\begin{array}{ccc}
\mathrm{Aff}(\mathbf{O}) & \xrightarrow{\quad \rho \quad} & \mathrm{Sp}(H^1(\mathbf{O}, \mathbb{Z})) \\
{\scriptstyle f \mapsto \varphi_A f \varphi_A^{-1}} \downarrow & & \downarrow {\scriptstyle \psi \mapsto (\varphi_A^{-1})^* \psi \varphi_A^*} \\
\mathrm{Aff}(A \cdot \mathbf{O}) & \xrightarrow[A \cdot \rho]{} & \mathrm{Sp}(H^1(A \cdot \mathbf{O}, \mathbb{Z}))
\end{array}$$

commutes.

For later use, we recall the following notion.

Definition 6.9 Let X be a path-connected, locally simply connected topological space, endowed with a base point $x \in X$, and let $R \subset S$ be rings.

a) An S-local system $\mathbb{V}$ is *defined over* R, if there is a R-local system $\mathbb{W} \subset \mathbb{V}$ such that $\mathbb{W} \otimes_R S \cong \mathbb{V}$.

b) Similarly, if V is an S-module and $\rho : \pi_1(X, x) \to \mathrm{Aut}(V)$ is a representation, then ρ is *defined over* R, if there is an R-submodule W of V and a representation $\sigma : \pi_1(X, x) \to \mathrm{Aut}(W)$, such that $\sigma \otimes_R S \cong \rho$.

6.3 Vector Bundles with a Flat Connection

In this section, let B always be a complex manifold. We complete the triptych by explaining the relationship between $\mathbb{C}$-local systems and holomorphic vector bundles that admit a certain first order-differential operator, called a connection. We restrict our attention to holomorphic connections, but point out that there is a C^∞-analogue (see e. g. [Voi02, Sect. 3.2.1]).

Definition 6.10 Let $\mathcal{V}$ be a holomorphic vector bundle on B. Denote the holomorphic cotangent sheaf of B by Ω_B^1. A holomorphic connection is a $\mathbb{C}$-linear map

$$\nabla : \mathcal{V} \to \Omega_B^1 \otimes_{\mathcal{O}_B} \mathcal{V},$$

which satisfies the Leibniz rule, i. e. for local sections f of $\mathcal{O}_B$ and s of $\mathcal{V}$, we have

$$\nabla(fs) = \mathrm{d}f \otimes s + f\nabla(s).$$

A morphism between vector bundles with connections $f : (\mathcal{V}, \nabla_1) \to (\mathcal{W}, \nabla_2)$ is a morphism between vector bundles $f : \mathcal{V} \to \mathcal{W}$ such that

$$(\mathrm{id} \otimes f) \circ \nabla_1 = \nabla_2 \circ f.$$

Remark 6.11 Let ∇ be a holomorphic connection on a vector bundle $\mathcal{V}$ of rank r over a complex manifold B. Let $U \subset B$ be a chart, where $\mathcal{V}$ can be trivialized, and choose a trivialization $\alpha : \mathcal{O}^r_{B\,|U} \to \mathcal{V}_{|U}$. Let e_i denote the i-th standard basis vector. Then

$$\nabla(\alpha(e_i)) = \sum_{j=1}^{r} \omega_{ji} \otimes \alpha(e_j).$$

We call the matrix

$$\Gamma_\alpha = \begin{pmatrix} \omega_{11} & \dots & \omega_{1r} \\ \vdots & & \vdots \\ \omega_{r1} & \dots & \omega_{rr} \end{pmatrix} \qquad \text{with} \quad \omega_{ij} \in \Omega^1_{B|U}$$

the matrix of the connection with respect to the chosen trivialization α. For any local section $(f_1, \dots, f_r)^T$ of $\mathcal{O}^r_{B\,|U}$, we have

$$\nabla_\alpha = \mathrm{id} \otimes \alpha^{-1} \circ \nabla \circ \alpha : (f_1, \dots, f_r)^T \mapsto (\mathrm{d}f_1, \dots, \mathrm{d}f_r)^T + \Gamma_\alpha \cdot (f_1, \dots, f_r)^T.$$

If another trivialization β is chosen, the base change is described by $A_{\alpha\beta} \in \mathrm{GL}(\mathcal{O}_{B\,|U})$, i. e. $\beta = \alpha \circ A_{\alpha\beta}$. Then for a local section $F = (f_1, \dots, f_r)^T$ of $\mathcal{O}^r_{B\,|U}$, we have

$$\begin{aligned} \nabla_\beta(F) &= A^{-1}_{\alpha\beta}(\mathrm{d}(A_{\alpha\beta}F)_1, \dots, \mathrm{d}(A_{\alpha\beta}F)_r)^T + A^{-1}_{\alpha\beta}\Gamma_\alpha A_{\alpha\beta} F \\ &= (\mathrm{d}f_1, \dots, \mathrm{d}f_r)^T + (A^{-1}_{\alpha\beta}\mathrm{d}A_{\alpha\beta} + A^{-1}_{\alpha\beta}\Gamma_\alpha A_{\alpha\beta})F. \end{aligned}$$

Therefore,

$$\Gamma_\beta = A^{-1}_{\alpha\beta}\mathrm{d}A_{\alpha\beta} + A^{-1}_{\alpha\beta}\Gamma_\alpha A_{\alpha\beta}.$$

Example 6.12 Let $\mathbb{V}$ be a $\mathbb{C}$-local system on B. Then there is a canonical connection associated with $\mathbb{V}$ on the vector bundle $\mathcal{V} = \mathbb{V} \otimes_{\mathbb{C}} \mathcal{O}_B$. Let U be an open set, where $\mathbb{V}$ can be trivialized, and let $s_1, \ldots, s_n$ be the basis of $\mathbb{V}(U)$ given by the trivialization. For a section $s = \sum_i f_i s_i \in \mathcal{V}(U)$, define

$$\nabla(s) = \sum_i \mathrm{d}f_i \otimes s_i.$$

This local definition is compatible with the coordinate changes, as they are given by locally constant matrices, hence it gives rise to a global map ∇. It follows from the definition that ∇ is $\mathbb{C}$-linear and satisfies the Leibniz rule, so ∇ is a connection on $\mathcal{V}$.

In particular, we have $\nabla(s_i) = 0$ for all $i - 1, \ldots, n$, and even $\mathrm{Ker}(\nabla) = \mathbb{V}$. For locally on a coordinate neighborhood U of B, where $\mathbb{V}$ can be trivialized, we can write the image $\nabla(s)$ of $s \in \mathcal{V}(U)$ as

$$\nabla(s) = \sum_i \mathrm{d}f_i \otimes s_i = \sum_{i,j} \frac{\partial f_i}{\partial z_j} \mathrm{d}z_j \otimes s_i$$

and the set $\{\mathrm{d}z_j \otimes s_i\}_{i,j}$ is a $\mathcal{O}_B(U)$-basis of $\Omega^1_B \otimes_{\mathcal{O}_B} \mathcal{V}(U)$. Therefore, $\nabla(s) = 0$ implies $\frac{\partial f_i}{\partial z_j} = 0$, and the f_i must be locally constant functions.

The example above hints at a relation between local systems and vector bundles with a connection. However, the latter do not arise from local systems unless they are flat, i. e. their curvature vanishes. To define the curvature of a vector bundle with connection $(\mathcal{V}, \nabla)$, consider the map

$$\tilde{\nabla} : \Omega^1_B \times \mathcal{V} \to \Omega^2_B \otimes_{\mathcal{O}_B} \mathcal{V}$$

which is given on local sections by

$$\tilde{\nabla}(\omega, \sigma) = \mathrm{d}\omega \otimes \sigma + \nabla(\sigma) \wedge \omega.$$

Here, $\Omega_B^p = \bigwedge^p \Omega_B$ $(p \in \mathbb{N})$. As

$$\begin{aligned}\tilde{\nabla}(f\omega,\sigma) &= \mathrm{d}(f\omega)\otimes\sigma + \nabla(\sigma)\wedge(f\omega)\\ &= (\mathrm{d}f\wedge\omega + f\mathrm{d}\omega)\otimes\sigma + f\nabla(\sigma)\wedge\omega\\ &= \mathrm{d}\omega\otimes f\sigma - \omega\wedge\mathrm{d}f\otimes\sigma - \omega\wedge f\nabla(\sigma)\\ &= \mathrm{d}\omega\otimes f\sigma + (\mathrm{d}f\otimes\sigma + f\nabla(\sigma))\wedge\omega\\ &= \mathrm{d}\omega\otimes f\sigma + \nabla(f\sigma)\wedge\omega = \tilde{\nabla}(\omega, f\sigma),\end{aligned}$$

for all $f \in \mathcal{O}_B$, we obtain a map $\nabla^{(1)} : \Omega_B^1 \otimes_{\mathcal{O}_B} \mathcal{V} \to \Omega_B^2 \otimes_{\mathcal{O}_B} \mathcal{V}$.

Definition 6.13 The curvature of the connection ∇ on $\mathcal{V}$ is defined as

$$R = \nabla^{(1)} \circ \nabla.$$

A connection is *flat* (or *integrable*), if $R = 0$.

We remark that R is an $\mathcal{O}_B$-linear map.

Example 6.14 Returning to Example 6.12, we show that the canonical connection on $\mathcal{V} = \mathbb{V} \otimes_{\mathbb{C}} \mathcal{O}_B$ is flat. For let again $s = \sum_i f_i \otimes s_i$ be a section in a local trivialization. Then

$$\begin{aligned}R(s) &= \nabla^{(1)}(\sum_i \mathrm{d}f_i \otimes s_i) = \sum_i \nabla^{(1)}(\mathrm{d}f_i \otimes s_i)\\ &= \sum_i (\mathrm{dd}f_i)\otimes s_i + \nabla(s_i)\wedge \mathrm{d}f_i = 0.\end{aligned}$$

We denote the canonical flat connection on $\mathcal{V}$ by $\nabla_{\mathbb{V}}$ or just ∇ if there is no ambiguity possible.

Proposition 6.15 ([Del70, Theorem 2.17])
Let B be a complex manifold. Then the functor

$$\mathbb{V} \mapsto (\mathbb{V} \otimes_{\mathbb{C}} \mathcal{O}_B, \nabla_{\mathbb{V}})$$

from the category of $\mathbb{C}$-local systems on B to the category of vector bundles on B equipped with a flat connection has a quasi-inverse

$$(\mathcal{V}, \nabla) \mapsto \mathrm{Ker}(\nabla).$$

7 Polarized Variations of Hodge Structures

We present the general definitions and properties of polarized variations of Hodge structures, with emphasis (and restriction) to the case of curves. As a reference for this theory, we recommend [Voi02], [Gri70], [Del71b].

7.1 Hodge Structures

Following P.Deligne [Del71b], we define

Definition 7.1 Let $V_{\mathbb{R}}$ be a finite-dimensional $\mathbb{R}$-vector space. A *real Hodge structure* of weight $k \in \mathbb{Z}$ on $V_{\mathbb{R}}$ is a decomposition of $V_{\mathbb{C}} = V_{\mathbb{R}} \otimes_{\mathbb{R}} \mathbb{C}$

$$V_{\mathbb{C}} = \bigoplus_{p+q=k} V^{p,q}$$

into complex subspaces $V^{p,q}$, such that we have $V^{q,p} = \overline{V^{p,q}}$, where complex conjugation acts on the second factor of $V_{\mathbb{C}}$.

Definition 7.2 A *Hodge structure* of weight $k \in \mathbb{Z}$ is a finitely generated abelian group V, together with a real Hodge structure of weight k on $V \otimes_{\mathbb{Z}} \mathbb{R}$. V is called the *integral lattice* of the Hodge structure.

Note that giving a real Hodge structure on $V_{\mathbb{R}}$ is the same as giving a decreasing filtration

$$\ldots \supset \mathrm{F}^p(V) \supset \mathrm{F}^{p+1}(V) \supset \ldots$$

of $V_{\mathbb{C}}$, which satisfies

$$V_{\mathbb{C}} = \mathrm{F}^p(V) \oplus \overline{\mathrm{F}^{k-p+1}(V)}$$

for each $p \in \mathbb{Z}$. We have

$$\mathrm{F}^p(V) = \bigoplus_{i \geq p} V^{i,k-i}$$

and

$$V^{p,q} = \mathrm{F}^p(V) \cap \overline{\mathrm{F}^q(V)}.$$

A Hodge structure will therefore be denoted either as a pair $(V, \{V^{p,q}\}_{p,q})$ or by its filtration $(V, \mathrm{F}^{\cdot}(V))$.

Definition 7.3 Let $(V, V^{p,q})$, $(W, W^{p,q})$ be Hodge structures of weight k and l. Let $(r,s) \in \mathbb{Z}^2$, such that $l = k + r + s$. A *morphism of Hodge structures* $f : (V, \mathrm{F}^{\cdot}) \to (W, \mathrm{F}^{\cdot})$ of bidegree (r,s) is a morphism $f : V \to W$ of groups, which is compatible with the Hodge decomposition, i. e. $f(V^{p,q}) \subset W^{p+r,q+s}$ (or equivalently with the filtration, i. e. $f(\mathrm{F}^p(V)) \subset \mathrm{F}^{p+r}(W)$).

Definition 7.4 Let $k \in \mathbb{Z}$.

a) A *polarization of a real Hodge structure* $(V_{\mathbb{R}}, V^{p,q})$ of weight k is a bilinear form $Q : V_{\mathbb{C}} \otimes V_{\mathbb{C}} \to \mathbb{C}$, symmetric for even k and alternating for odd k, such that the generalized Riemann relations are satisfied, i. e.

$$Q(V^{p,q}, V^{p',q'}) = 0 \qquad \text{unless } p' = q,\ q' = p \tag{7.1}$$

$$i^{p-q} Q(u, \overline{u}) > 0 \qquad \text{for } p + q = k \text{ and } u \in V^{p,q} \setminus \{0\}. \tag{7.2}$$

These relations can be reinterpreted in the following way. Introduce the Weil operator $C : V_{\mathbb{C}} \to V_{\mathbb{C}}$; for $u \in V^{p,q}$, set $C(u) = i^{p-q} u$. Then the generalized Riemann relations are equivalent to the assertion that $(u, v) \mapsto Q(C(u), \overline{v})$ is a positive definite hermitian form on V, for which the Hodge decomposition is orthogonal.

b) In addition to that, a *polarization of a Hodge structure* $(V, V^{p,q})$ of weight k is required to take only integer values on the underlying lattice V.

c) A *morphism of polarizable Hodge structures* is a morphism of the underlying (unpolarized) Hodge structures. A *morphism of polarized Hodge structures* is required to respect the polarization, i. e. the one in the range is the pullback of the one in the image.

Starting from a finite-dimensional k-vector space V instead of a finitely generated abelian group, where k is a subfield of $\mathbb{R}$, one also defines k-Hodge structures in an analogous way. A polarization of a k-Hodge structure is required to take values in k on V.

Example 7.5 The standard example of a Hodge structure of weight k is the one given by the Hodge decomposition of $H^k(X, \mathbb{Z})$ of a compact Kähler manifold X (compare with Remark 4.9). If the class of the Kähler form $[\omega] \in H^2(X, \mathbb{Z})$, then we have a polarization on the sub-Hodge structure given by the Lefschetz decomposition (see [Voi02, Ch. 7]). The existence of an integral Kähler class $[\omega]$ has strong implications: The manifold X is a projective variety.

Weight 1

Let us consider the polarized Hodge structures, which model the Hodge decomposition of the first cohomology of a compact Riemann surface. We call a Hodge structure *pure*, if $V^{p,q} = 0$, when $p < 0$ or $q < 0$. As we show in the following, a pure, polarized Hodge structure of weight 1 is nothing else, but a polarized abelian variety.

Recall that a polarized abelian variety is a complex torus $A = W/\Lambda$ together with the first Chern class $E = c_1(L) \in H^2(A, \mathbb{Z})$ of a positive definite line bundle L on A. E can be interpreted in a canonical way as an $\mathbb{R}$-bilinear, alternating form on W that takes integral values on Λ, and in addition satisfies

$$E(iu, iv) = E(u, v) \quad \text{and} \quad E(iu', u') > 0$$

for all $u, v \in W$ and $u' \in W \setminus \{0\}$. The first equation says that the assignment $(u, v) \mapsto H(u, v) = E(iu, v) + iE(u, v)$ defines a hermitian form on W. The inequality asserts that H is positive definite. We can choose a symplectic basis $\{\lambda_i, \mu_i\}_{i=1}^g$ for $\Lambda \subset W$, such that E with respect to this basis has the matrix

$$\begin{pmatrix} 0 & D \\ -D & 0 \end{pmatrix}$$

where $D = \mathrm{diag}(d_1, \ldots, d_g)$ such that $d_i \mid d_{i+1}$. D is called the type of the polarization. A polarization of type $D = I_g$ is called *principal.* A morphism of polarized abelian varieties (A, E), (A', E') is a holomorpic group homomorphism $f : A \to A'$ such that $f^*E' = E$.

Remark 7.6 The category of polarized abelian varieties $(A = W/\Lambda, E)$ and the category of pure, polarized Hodge structures $(V, V^{p,q})$ of weight 1 are equivalent.

Given a Hodge structure with integral lattice V and polarization Q, let $\pi : V_{\mathbb{C}} \to V^{0,1}$ be the projection, and let $\psi : V^{0,1} \to V_{\mathbb{R}} \subset V_{\mathbb{C}}$ be the section defined by $v \mapsto v + \overline{v}$. Let $A = V^{0,1}/\pi(V)$. Then A is a complex torus, since the map $V \otimes_{\mathbb{Z}} \mathbb{R} \hookrightarrow V_{\mathbb{C}} \xrightarrow{\pi} V^{0,1}$ is an $\mathbb{R}$-linear isomorphism of $\mathbb{R}$-vector space, and V is a lattice in $V \otimes_{\mathbb{Z}} \mathbb{R}$. For $u, v \in V^{0,1}$, define $E(u, v) = -Q(\psi(u), \psi(v))$. It follows from the generalized Riemann relations that E is the first Chern class of a line bundle on A, thus a polarization of A. Note that the hermitian form corresponding to E is just $Q(C(\cdot), \bar{\cdot})$.

Conversely, given a polarized abelian variety $A = W/\Lambda$ of dimension g, consider the Hodge decomposition of its first cohomology

$$H^1(A, \mathbb{C}) = H^{1,0} \oplus H^{0,1}.$$

The underlying lattice is the $\mathbb{Z}$-dual of Λ, since $\Lambda \cong H_1(A, \mathbb{Z})$. The polarization of A, i. e. the integral Kähler class $E = [\omega]$ gives a polarization of the Hodge structure

$$Q(\alpha, \beta) = \int_A \omega^{g-1} \wedge \alpha \wedge \beta$$

(see [Voi02, Sect.7.2.2]). If we take the dual Hodge structure, then this defines a quasi-inverse to the functor from polarized Hodge structures to polarized abelian varieties.

The Polarized Period Domain in Weight 1

One can consider the collection of all possible Hodge structures of weight k that can be put on a finitely generated abelian group with fixed polarization, and assemble them into a space. This is Griffiths' period domain [Gri68a], [Sch73, Sect. 3].

If we restrict to pure, polarized Hodge structures of weight 1, then in view of Remark 7.6, the Siegel upper-half space

$$\mathfrak{H}_g = \left\{ Z \in \mathbb{C}^{g\times g} \mid Z^T = Z \text{ and } \operatorname{Im}(Z) > 0 \right\}$$

is the classifying space. Since it is sometimes preferable to have a more abstract description at hand, we describe the period domain from an abstract point of view, and make the above correspondence explicit in the following. A reference for the Siegel upper half space is [BL04, Chap. 8].

Definition 7.7 Let V be a free abelian group of rank $2g$, $V_{\mathbb{R}} = V \otimes_{\mathbb{Z}} \mathbb{R}$, and $V_{\mathbb{C}} = V_{\mathbb{R}} \otimes_{\mathbb{R}} \mathbb{C}$ and let $Q : V_{\mathbb{C}} \otimes V_{\mathbb{C}} \to \mathbb{C}$ be a non-degenerate, alternating form, taking integral values on V. Let $\mathfrak{D} = \mathfrak{D}(V,Q) \subset \operatorname{Grass}(g, V_{\mathbb{C}})$ be the subset of the g-dimensional subspaces W of $V_{\mathbb{C}}$ which obey to the Riemann bilinear relations

$$Q(W,W) = 0 \tag{7.3}$$

$$iQ(w,\overline{w}) > 0 \qquad \text{for all } \; w \in W \tag{7.4}$$

$\mathfrak{D}$ is called *period domain of pure Hodge structures of weight* 1 *on* V*, polarized by* Q.

Definition 7.8 Let V and Q be as in Definition 7.7. Let G_Q be the linear algebraic group over $\mathbb{Z}$ of transformations, which are orthogonal for Q. Explicitly, for a ring R

$$G_Q(R) = \left\{ g \in \operatorname{GL}(V \otimes_{\mathbb{Z}} R) \mid Q(gu, gv) = Q(u,v) \; \text{ for all } \; u, v \in V \otimes_{\mathbb{Z}} R \right\}.$$

Remark 7.9 Let $\check{\mathfrak{D}} = \check{\mathfrak{D}}(V,Q)$ be the subset of points in $\operatorname{Grass}(g, V_{\mathbb{C}})$, which only satisfy (7.3).

a) $G_Q(\mathbb{C})$ acts transitively on $\check{\mathfrak{D}}$ with closed isotropy group; in particular, $\check{\mathfrak{D}}$ is a non-singular subvariety of $\operatorname{Grass}(g, V_{\mathbb{C}})$.

b) $\mathfrak{D}$ is an open set (in the Hausdorff topology) of $\check{\mathfrak{D}}$.

c) $G_Q(\mathbb{R})$ acts transitively on $\mathfrak{D}$ with compact isotropy group.

d) Any discrete subgroup of $G_Q(\mathbb{R})$ acts properly discontinuously and holomorphically on $\mathfrak{D}$.

PROOF: a) Transitivity of the actions follows from arguments from linear algebra. G_Q is closed in GL_{2g}, which acts on $\mathrm{Grass}(g, V)$ with closed isotropy groups; this implies all the assertions of a). b) is clear, since (7.4) is an open condition. For c) and d) we refer to [BL04, Sect. 8.2], where the assertions are proved for Siegel's upper half space. ■

Note that $G_Q = \mathrm{Sp}$ if the polarization is principal.

Remark 7.10 As described in [Gri68a, Proposition 1.24], $\mathfrak{D}(V, Q)$ is analytically isomorphic to $\mathfrak{H}_g$, the classifying space for polarized abelian varieties with a fixed symplectic basis. We sketch this isomorphism. Let $\{\gamma_i\}_{i=1}^{2g}$ be a symplectic basis of V, such that Q has the matrix

$$A_Q = \begin{pmatrix} 0 & D \\ -D & 0 \end{pmatrix}$$

with $D = \mathrm{diag}(d_1, \ldots, d_g) \in \mathbb{Z}^{g \times g}$, and let $W \in \mathfrak{D}$. Choose a basis $\omega_1, \ldots, \omega_g$ of W. Then

$$\omega_i = \sum_{j=1}^{2g} \sigma_{ij} \gamma_j ,$$

and we let

$$\Omega = (\sigma_{ij})_{i=1,\ldots,g, j=1,\ldots,2g} \in \mathbb{C}^{g \times 2g} .$$

Ω is called *period matrix*. $(\Omega^T | \overline{\Omega}^T)$ is the base change from $\{\omega_i, \overline{\omega_i}\}_{i=1}^{g}$ to $\{\gamma_j\}_{j=1}^{2g}$. In particular, it follows from (7.3) and (7.4) that

$$\Omega A_Q \Omega^T = 0 \quad \text{and} \quad i \Omega A_Q \overline{\Omega}^T \text{ is positive definite.}$$

There is an action from the right on such period matrices Ω by $\mathrm{GL}_g(\mathbb{C})$ by $\Omega \cdot A = A^T \Omega$, induced by the base change

$$\begin{pmatrix} A & 0 \\ 0 & \bar{A} \end{pmatrix} .$$

Writing $\Omega = (E|F)$ with $E, F \in \mathbb{C}^{g\times g}$, we have that F and E are both regular, for

$$i(ED\overline{F}^T - FD\overline{E}^T)$$

is positive definite. So $\Omega\cdot\mathrm{GL}(W) \ni (Z|D^{-1})$ with $Z = D^{-1}F^{-1}E \in \mathbb{C}^{g\times g}$. Sending W to $Z \in \mathfrak{H}_g$ defines the isomorphism.

There is an action on the left by $G_Q(\mathbb{R})$ by $M \cdot \Omega = \Omega M^T$, which is a change of the symplectic basis. In view of the above, we can identify this action on $\mathfrak{H}_g$. For simplicity assume that $D = I$, i. e. that Q is a principal polarization. Then the action is given by

$$G_Q(\mathbb{R}) \times \mathfrak{H}_g \to \mathfrak{H}_g, \quad (M, Z) \mapsto M(Z) = (cZ + d)^{-1}(aZ + b)$$

where $M = \left(\begin{smallmatrix} a & b \\ c & d \end{smallmatrix}\right)$.

Example 7.11 Let X be a compact topological surface of genus $g \geq 1$. Let i be the intersection form on $H_1(X, \mathbb{Z})$, and choose a symplectic basis $\{a_i, b_i\}_{i=1}^g$. Any complex structure on X determines a g-dimensional subvector space $\Omega_X^1(X) \subset H^1(X, \mathbb{C})$, the $(1,0)$-part of the Hodge decomposition. Thus $W = \Omega_X^1(X)$ determines a point in $\mathfrak{D}(H^1(X, \mathbb{Z}), i^*)$. The name period domain stems from choosing a basis $\omega_1, \ldots, \omega_g$ of $\Omega_X^1(X)$ and considering the matrix of periods

$$\Omega = \begin{pmatrix} \int_{a_1} \omega_1 & \cdots & \int_{a_g} \omega_1 & \int_{b_1} \omega_1 & \cdots & \int_{b_g} \omega_1 \\ \vdots & & \vdots & \vdots & & \vdots \\ \int_{a_1} \omega_g & \cdots & \int_{a_g} \omega_g & \int_{b_1} \omega_g & \cdots & \int_{b_g} \omega_g \end{pmatrix}.$$

7.2 Variations of Hodge Structures

Definition 7.12 Let B be a connected complex manifold, and let $k \in \mathbb{Z}$.

a) A *real variation of Hodge structures* ($\mathbb{R}$-VHS) of weight k on B is given by a local system $\mathbb{V}_\mathbb{R}$ of stalk $V_\mathbb{R}$, where $V_\mathbb{R}$ is a finite-dimensional $\mathbb{R}$-vector space, together with a decreasing filtration $\mathrm{F}^{\cdot}$ of the holomorphic vector bundle $\mathcal{V} = \mathbb{V}_\mathbb{R} \otimes_\mathbb{R} \mathcal{O}_B$ by holomorphic subbundles F^p. These should satisfy the following conditions:

(i) The filtration $\mathrm{F}^{\cdot}$ satisfies Griffiths' transversality condition with respect to the flat connection $\nabla : \mathcal{V} \to \mathcal{V} \otimes_{\mathcal{O}_B} \Omega^1_B$ associated with $\mathbb{V}_{\mathbb{C}} = \mathbb{V}_{\mathbb{R}} \otimes_{\mathbb{R}} \mathbb{C}$ (compare Example 6.12), i. e.

$$\nabla(\mathrm{F}^p(\mathcal{V})) \subset \mathrm{F}^{p-1}(\mathcal{V}) \otimes_{\mathcal{O}_B} \Omega^1_B.$$

(ii) For every $b \in B$, the fiber $\mathrm{F}^{\cdot}(\mathcal{V})_b$ over b of the filtration induces a real Hodge structure of weigth k on the stalk $(\mathbb{V}_{\mathbb{R}})_b \cong V_{\mathbb{R}}$.

b) A *variation of Hodge structures (VHS) of weight* k on B is a local system $\mathbb{V}_{\mathbb{Z}}$ of stalk V, where V is a finitely generated, free abelian group, together with an $\mathbb{R}$-VHS of weight k on $\mathbb{V}_{\mathbb{Z}} \otimes_{\mathbb{Z}} \mathbb{R}$.

c) Let $L \subset \mathbb{R}$ be a subfield. An *L-VHS of weight* k on B is an L-local system $\mathbb{V}_L$, together with an $\mathbb{R}$-VHS of weight k on $\mathbb{V}_L \otimes_L \mathbb{R}$.

d) A *morphism of VHS* $f : (\mathbb{V}, \mathrm{F}^{\cdot}) \to (\mathbb{W}, \mathrm{F}^{\cdot})$ is a morphism of the underlying local systems, which is compatible with the filtrations (as in 7.3).

Note that for $k = 1$, the transversality condition is vacuous.

Definition 7.13 Let B be a connected complex manifold. A *complex variation* of weight k consists of a $\mathbb{C}$-local system $\mathbb{V}_{\mathbb{C}}$ and a decomposition of $\mathcal{V} = \mathbb{V}_{\mathbb{C}} \otimes_{\mathbb{C}} \mathrm{C}^{\infty}_B$ into C^{∞}-subbundles $\mathcal{V}^{p,q}$

$$\mathcal{V} = \bigoplus_{p+q=k} \mathcal{V}^{p,q}$$

such that the C^{∞}-flat connection $D : \mathcal{V} \to \mathcal{V} \otimes_{\mathrm{C}^{\infty}_B} \mathcal{A}^1_B$ associated with $\mathbb{V}_{\mathbb{C}}$ sends sections of $\mathcal{V}^{p,q}$ into

$$(\mathcal{V}^{p+1,q-1} \otimes \mathcal{A}^{0,1}_B) \oplus (\mathcal{V}^{p,q} \otimes \mathcal{A}^1_B) \oplus (\mathcal{V}^{p-1,q+1} \otimes \mathcal{A}^{1,0}_B).$$

We remark that a VHS of weight k induces a complex variation of weight k on $\mathbb{V}_{\mathbb{C}}$. Also, pure variations of Hodge structures are defined analogously to pure Hodge structures.

One notation that we will make use of in the following. If $\mathbb{V}$ is an R-local system, and if A is an R-algebra, we denote $\mathbb{V}_A$ the local system $\mathbb{V} \otimes_R A$ obtained by tensoring with the constant sheaf R.

Definition 7.14 A *polarization of an* $\mathbb{R}$-*VHS* $(\mathbb{V}_{\mathbb{R}}, F^{\cdot})$ of weight $k \in \mathbb{Z}$ on B is a $\mathbb{C}$-bilinear, locally constant map $Q : \mathbb{V}_{\mathbb{C}} \otimes \mathbb{V}_{\mathbb{C}} \to \mathbb{C}_B$, which induces a polarization of the Hodge structure of the stalk $(\mathbb{V}_{\mathbb{R}})_b$ for each $b \in B$.

For a *polarization of a VHS* $(\mathbb{V}_{\mathbb{Z}}, F^{\cdot})$, we require in addition that Q restricted to $\mathbb{V}_{\mathbb{Z}}$ take values only in $\mathbb{Z}$.

Polarized variations of Hodge structures will be abbreviated by pVHS.

In particular, a pure pVHS of weight 1 on B will be denoted as a triple $(\mathbb{V}, \mathcal{V}^{1,0}, Q)$, where $\mathbb{V}$ is the local system, $\mathcal{V}^{1,0}$ is the only relevant step of the filtration of $\mathcal{V} = \mathbb{V} \otimes_{\mathbb{Z}} \mathcal{O}_B$ and Q is the polarization. A sub-pVHS $(\mathbb{L}, \mathcal{L}^{1,0})$ of $(\mathbb{V}, \mathcal{V}^{1,0}, Q)$ is a monomorphism of VHS $(\mathbb{L}, \mathcal{L}^{1,0}) \to (\mathbb{V}, \mathcal{V}^{1,0})$ such that $(\mathbb{L}, \mathcal{L}^{1,0})$ is polarized by the bilinear form induced by Q via pullback.

One of the cornerstones of the study of pVHS is P. Deligne's result on semisimplicity, which we state in the following form.

Proposition 7.15 ([Del87, Proposition 1.13])
Let B be a connected complex manifold inside a connected, compact complex manifold $\overline{B}$, such that $\overline{B} \setminus B$ is a complex submanifold of $\overline{B}$. Let $b \in B$ be a base point, and let $(\mathbb{V}, F^{\cdot}, Q)$ be a polarized VHS of weight k. Then

a) The local system $\mathbb{V}_{\mathbb{C}} = \mathbb{V} \otimes_{\mathbb{Z}} \mathbb{C}$, or equivalently the action of $\pi_1(B, b)$ on $(\mathbb{V}_{\mathbb{C}})_b$, is completely reducible,

$$\mathbb{V}_{\mathbb{C}} = \bigoplus_{i \in I} \mathbb{L}_i \otimes W_i \ ,$$

where the local systems $\mathbb{L}_i$ are irreducible and mutually non-isomorphic, and W_i are $\mathbb{C}$-vector spaces.

b) Every $\mathbb{L}_i$ carries a polarized complex variation of weight k, and the polarized complex variation induced on $\mathbb{V}$ by the ones on $\mathbb{L}_i$ and W_i is the complex variation of the VHS.

7.3 Variations of Hodge Structures and Families of Curves

We shortly summarize how a family of curves gives rise to a polarized VHS.

Proposition 7.16 *Let B be a complex manifold, and let $\phi : \mathcal{X} \to B$ be a family of curves of genus g. Then ϕ defines canonically a VHS on B in the following way: The underlying local system is $\mathbb{V}_{\mathbb{Z}} = R^1\phi_*(\mathbb{Z}_{\mathcal{X}})$ and the Hodge filtrations on the fibers $\mathcal{X}_b$ can be glued together to give a holomorphically varying subbundle $\mathcal{V}^{1,0}$ of the vector bundle $\mathcal{V} = \mathbb{V}_{\mathbb{Z}} \otimes_{\mathbb{Z}} \mathcal{O}_B$. Explicitly,*

$$\mathcal{V}^{1,0} = \phi_*\Omega^1_{\mathcal{X}\,/\,B}$$

the direct image of the sheaf of relative differentials.

The VHS is automatically polarized by the unique global section of $R^2\phi_*(\mathbb{Z}_{\mathcal{X}})$ given by the complex structure.

In this case, the holomorphic connection ∇ associated with $\mathbb{V}_{\mathbb{C}} = \mathbb{V}_{\mathbb{Z}} \otimes_{\mathbb{Z}} \mathbb{C}$ has been called *Gauß-Manin connection* by Grothendieck.

PROOF: One way to proof this proposition is to use the holomorphicity of Griffiths period map (see below). Another more algebraic approach is sketched in [Del71b]: Since ϕ is proper, we have

$$\mathcal{V} = R^1\phi_*(\mathbb{Z}) \otimes \mathcal{O}_B \cong R^1\phi_*(\phi^{-1}\,\mathcal{O}_B).$$

The relative local system $\phi^{-1}\,\mathcal{O}_B$ has a resolution by the relative holomorphic de Rham complex

$$\Omega^{\cdot}_{\mathcal{X}\,/\,B} = (0 \longrightarrow \mathcal{O}_B \xrightarrow{\ \mathrm{d}\ } \Omega^1_{\mathcal{X}\,/\,B} \longrightarrow 0).$$

This implies that $R^1\phi_*(\phi^{-1}\,\mathcal{O}_B)$ is isomorphic to $R^1\phi_*(\Omega^{\cdot}_{\mathcal{X}\,/\,B})$, the hypercohomology of the complex $\Omega^{\cdot}_{\mathcal{X}\,/\,B}$. This leads to a spectral sequence

$$E_1^{p,q} = R^q\phi_*(\Omega^p_{\mathcal{X}\,/\,B}) \Rightarrow R^1\phi_*(\phi^{-1}\,\mathcal{O}_B) \cong \mathcal{V}\,.$$

Now, Deligne argues that $E_1^{p,q}$ is locally free, that the spectral sequence degenerates at E_1, and that its formation comutes with base change. It defines a filtration on $\mathcal{V}$

$$R^0\phi_*(\Omega^1_{\mathfrak{X}\,/\,B}) = \phi_*(\Omega^1_{\mathfrak{X}\,/\,B}) \subset \mathcal{V}$$

and this filtration induces the Hodge filtration on each fiber by base change. ∎

The Period Mapping

Let B be a connected complex manifold, and let $(\mathbb{V}, \mathcal{V}^{1,0}, Q)$ be a pure, polarized VHS of weight 1 on B. Fix a base point $b \in B$ and a universal covering $u : \tilde{B} \to B$. On $\tilde{B}$, the pullbacks of the local systems $\mathbb{V}$ and $\mathbb{V}_{\mathbb{C}}$ are isomorphic to the constant sheaf of stalk $\mathbb{V}_b$, respectively $(\mathbb{V}_{\mathbb{C}})_b$. The VHS pulls back to a VHS $(u^{-1}\,\mathbb{V}, u^*\,\mathcal{V}^{1,0}, u^*Q)$ on $\tilde{B}$, polarized by u^*Q.

Let $\tilde{b} \in u^{-1}(b)$, and let $\varphi_{\tilde{b}} : (u^{-1}\,\mathbb{V})_{\tilde{b}} \to \mathbb{V}_b$ be the canonical isomorphism. Let $z \in \tilde{B}$. There is a unique way of identifying germs in $(u^{-1}\,\mathbb{V})_z$ with germs in $(u^{-1}\,\mathbb{V})_{\tilde{b}}$ by continuation (along any path c connecting z to $\tilde{b}$); let $\Phi_{z,\tilde{b}} : (u^{-1}\,\mathbb{V})_z \to (u^{-1}\,\mathbb{V})_{\tilde{b}}$ be the induced isomorphism. The holomorphic bundle $\mathcal{V}^{1,0}$ on B pulls back to a holomorphic vector bundle $u^*\,\mathcal{V}^{1,0}$ on $\tilde{B}$ and singles out a subspace $(u^*\,\mathcal{V}^{1,0})_z \subset (u^{-1}\,\mathbb{V}_{\mathbb{C}})_z$. We let W_z be the image of $(u^*\,\mathcal{V}^{1,0})_z$ under $\varphi_{\tilde{b}} \circ \Phi_{z,\tilde{b}}$ inside $(\mathbb{V}_{\mathbb{C}})_b$. Since the polarization Q is locally constant, W_z obeys to the Riemann relations with respect to Q. We define the *period mapping* by

$$p : \tilde{B} \to \mathfrak{D}(\mathbb{V}_b, Q_b)\ , \quad z \mapsto W_z .$$

Remark 7.17 Let $p : \tilde{B} \to \mathfrak{D}(\mathbb{V}_b, Q_b)$ be the period mapping associated with a pure, polarized VHS of weight 1 on the complex connected manifold B (with fixed base point $b \in B$). Moreover, let $\rho : \pi_1(B, b) \to G_{Q_b}(\mathbb{Z})$ be the monodromy representation associated with the local system $\mathbb{V}$ by Proposition 6.4. Then

a) p is holomorphic.

b) p is equivariant with respect to the action of $\gamma \in \pi_1(B, b)$ on $\tilde{B}$ by deck transformations and $\rho(\gamma)$ on $\mathfrak{D}(\mathbb{V}_b, Q_b)$.

PROOF: a) was shown by Griffiths [Gri68b, Theorem 1.27]. b) Note that $\tilde{B}$ can be identified with pairs (x, α), where $x \in B$ and α is a homotopy class of paths from x to $\tilde{b}$. γ acts on $z = (x, \alpha)$ by $\gamma \cdot z = (x, \gamma\alpha)$ from the left. Therefore $p(\gamma z)$ is the subspace of $(\mathbb{V}_{\mathbb{C}})_b$ given as the image of $(u^* \mathcal{V}^{1,0})_z$ under

$$\varphi_{\gamma \cdot \tilde{b}} \circ \Phi_{z, \gamma \cdot \tilde{b}} = \varphi_{\gamma \cdot \tilde{b}} \circ \Phi_{\tilde{b}, \gamma \cdot \tilde{b}} \circ \Phi_{z, \tilde{b}} \, .$$

Since

$$\varphi_{\gamma \cdot \tilde{b}} \circ \Phi_{\tilde{b}, \gamma \cdot \tilde{b}} \circ \varphi_{\tilde{b}}^{-1} = \rho(\gamma) : \mathbb{V}_b \to \mathbb{V}_b,$$

the claim follows. ■

By abuse of language, we will also call the induced map

$$\overline{p} : B \to \rho(\pi_1(B, b)) \backslash \mathfrak{D}(\mathbb{V}_b, Q_b)$$

a period mapping.

Remark 7.18 Let $\phi : \mathcal{X} \to B$ be a family of curves of genus g on a connected complex manifold B, let $b \in B$, and let p be the period mapping associated with the polarized VHS $(\mathbb{V}, \mathcal{V}^{1,0}, Q)$ given by Proposition 7.16 and the base point b. The induced mapping

$$\overline{p} : B \to \rho(\pi_1(B, b)) \backslash \mathfrak{D}(\mathbb{V}_b, Q_b) \to \mathcal{A}_g$$

maps each point $x \in B$ to the isomorphism class of the Jacobian $\mathrm{Jac}(\mathcal{X}_x)$ of the fiber $\mathcal{X}_x$.

In the following, we describe the period mapping in the case of a Teichmüller curve. We only discuss the case of a pVHS of rank two as it will be important for later applications. Let (X, ω) be a Veech surface of genus g, and let $\Delta = \Delta(X, \omega^{\otimes 2}) \subset \mathcal{T}_g = \mathcal{T}(X)$ be the Teichmüller disk associated with (X, ω). Recall from Section 2.3 that we can parametrize a point in Δ by an element $\tau \in \mathbb{H} = \mathrm{SO}(2) \backslash \mathrm{SL}_2(\mathbb{R})$ and write $x_\tau = (X_\tau, m_\tau)$ for a point in Δ. The pullback of the universal family $\mathcal{X}_{\mathrm{univ}} \to \mathcal{T}_g$ to Δ is a family of curves $f : \mathcal{X} \to \Delta$.

Now assume that we are given a subgroup $\Gamma \leqslant \mathrm{Aff}(X, \omega)$ of finite index and a Γ-invariant subspace U of $H^1(X, \mathbb{Z})$ of rank 2. U corresponds to a sub-local system $\mathbb{L}$ of $R^1 f_*(\mathbb{Z}_{\mathcal{X}})$. Suppose that $\mathbb{L}$ carries a sub-pVHS of the pVHS on $R^1 f_*(\mathbb{Z}_{\mathcal{X}})$. The (1,0)-part of $\mathbb{L}$ is a holomorphic line bundle on Δ. Thus it can be globally trivialized (see [For81, Theorem 30.4]). Let it be generated by a global section $\omega \in H^0(\Delta, f_*\Omega_{\mathcal{X}/\Delta})$. Choose a symplectic basis $\{a, b\}$ of $U^* \leqslant H_1(X, \mathbb{Z})$ and modify ω by a change of basis in $H^0(\Delta, \mathcal{O}^\times)$ in such a way that

$$\int_{(m_{x_\tau})_* a} \omega_{x_\tau} \in \mathbb{H} \quad \text{and} \quad \int_{(m_\tau)_* b} \omega_{x_\tau} = 1,$$

where $x_\tau = (X_\tau, m_\tau : X \to X_\tau)$ runs over all points in Δ and ω_{x_τ} is the element of the fiber of $\mathbb{L}^{1,0}$ over x_τ. Precomposing with $\mathbb{H} \to \Delta$, $\tau \mapsto x_\tau$ yields a map

$$p : \mathbb{H} \to \mathbb{H}, \quad \tau \mapsto \int_{(m_\tau)_* a} \omega_{x_\tau}$$

which is an explicit version of the period map associated with the pVHS $\mathbb{L}$. Here Γ acts on the left-hand side by its action on Δ as a subgroup of $\mathrm{Aff}(X, \omega)$ and on the right-hand side by the monodromy representation. Explicitly, if $\gamma \in \Gamma$ acts on U as $\left(\begin{smallmatrix} A & B \\ C & D \end{smallmatrix}\right) \in \mathrm{SL}_2(\mathbb{Z})$ w. r. t. the (dual) basis a^*, b^* of U, then following Remark 7.10, this action on $\mathbb{H}$ is given by the usual Möbius transformation action

$$z \mapsto \frac{Az + B}{Cz + D}.$$

Lemma 7.19 *The map p constructed above is equivariant w. r. t. the action of Γ on source and target.*

PROOF: We consider the projective tuple

$$(z_a : z_b) = \left(\int_{(m_\tau)_* a} \omega_{x_\tau} : \int_{(m_\tau)_* b} \omega_{x_\tau} \right) = (p(\tau) : 1).$$

Let $\gamma \in \Gamma$. Then $\gamma \cdot x_\tau = (X_\tau, m_\tau \circ \gamma^{-1})$. Therefore,

$$(p(\gamma \cdot x_\tau) : 1) = (\int_{(m_\tau \circ \gamma^{-1})_* a} \omega_{\gamma \cdot x_\tau} : \int_{(m_\tau \circ \gamma^{-1})_* b} \omega_{\gamma \cdot x_\tau}).$$

Note that $\omega_{\gamma\cdot x_\tau}$ is proportional to ω_{x_τ}, and that γ acts on U^* as $\left(\begin{smallmatrix} D & -C \\ -B & A \end{smallmatrix}\right)$ w. r. t. to the basis a, b. Hence,

$$\gamma_*^{-1} a = Aa + Bb \quad \text{and} \quad \gamma_*^{-1} b = Ca + Db,$$

and we obtain

$$(p(\gamma \cdot x_\tau) : 1) = (Az_a + Bz_b : Cz_a + Dz_b) = \left(\frac{Az_a + B}{Cz_a + D} : 1\right),$$

as $z_b = 1$. Thus

$$p(\gamma \cdot x_\tau) = \frac{Ap(x_\tau) + B}{Cp(x_\tau) + D} = \rho(\gamma) p(x_\tau),$$

which proves the claim. ∎

Deligne Extension

We describe a canonical extension of a vector bundle with flat connection over singular points. We restrict ourselves to the case where the base is a pre-compact Riemann surface. A discussion of the general case can be found e. g. in [Del70], [Sch73].

For the rest of this section, let $\mathcal{C}$ be a pre-compact Riemann surface sitting in a compact surface $\overline{\mathcal{C}}$, let $i : \mathcal{C} \to \overline{\mathcal{C}}$ be the inclusion, and let $S = \overline{\mathcal{C}} \setminus \mathcal{C}$ be the finite set of cusps.

Definition 7.20 Let $\mathbb{V}$ be a $\mathbb{C}$-local system on $\mathbb{D}^* = \mathbb{D} \setminus \{0\}$, let $x \in \mathbb{D}^*$, and let γ denote a generator of $\pi_1(\mathbb{D}^*, x)$. Then $\mathbb{V}$ is said to have *unipotent monodromy*, if the image of γ under the representation $\rho : \pi_1(\mathbb{D}^*, x) \to \mathrm{GL}(\mathbb{V}_x)$ associated with $\mathbb{V}$ is a unipotent transformation.

Note that this definition is independent of the choice of the generator γ and the base point x.

In fact, the local systems that we are interested in are almost unipotent. P. Griffiths [Gri70, Theorem 3.1] recounts four ways of proving the next proposition.

Proposition 7.21 *Let $\phi : \mathcal{X} \to \mathcal{C}$ be a family of curves of genus g. If $s \in S$, and $T = \rho(\gamma)$ is the monodromy transformation of a small loop about s, then T is quasi-unipotent, i. e. there exist N, $M \in \mathbb{N}$ such that*

$$(T^N - 1)^M = 0.$$

We define the sheaf of 1-forms with log-singularities at S to be the subsheaf $\Omega^1_{\overline{\mathcal{C}}}(S)$ of $i_* \Omega^1_{\mathcal{C}}$ consisting of the local sections $\omega \in i_* \Omega^1_{\mathcal{C}}(U)$ on $U \subset \overline{\mathcal{C}}$ such that ω and $d\omega$ have a pole of order at most 1 in $S \cap U$. Note that since we are in the one-dimensional case,

$$\Omega^1_{\overline{\mathcal{C}}}(S) = \Big\{\omega \in \mathcal{M}^{(1)}_{\overline{\mathcal{C}}}(\overline{\mathcal{C}}) \mid \operatorname{div}(\omega) + \sum_{s \in S} s \geq 0\Big\} \cup \{0\}.$$

Definition 7.22 Let $c \in \mathcal{C}$ and $\mathbb{V}$ a $\mathbb{C}$-local system on $\mathcal{C}$. Let ∇ be the flat, holomorphic connection associated with $\mathbb{V}$ by Proposition 6.15. We make the following definitions.

a) $\mathbb{V}$ has *unipotent monodromy about a cusp* $s \in S$ if $\mathbb{V}$ restricted to a punctured neighborhood of s has unipotent monodromy.

b) $\mathbb{V}$ has *unipotent monodromy about the cusps* if a) holds for all $s \in S$.

c) ∇ is *meromorphic* at $s \in S$, if the coefficients of the matrix Γ of ∇ in a local trivialization are meromorphic.

d) ∇ is *regular* at $s \in S$, if ∇ is meromorphic at s and the matrix Γ in c) has at most poles of order 1 at s, i. e. entries in the sheaf $\Omega^1_{\overline{\mathcal{C}}}(S)$.

Note that by virtue of Remark 6.11, the definition in point c) above is independent of the choice of a local trivialization.

Lemma 7.23 *Let $\mathcal{V}$ be a holomorphic vector bundle of rank r on $\mathbb{D}$, let $\mathbb{D}^* = \mathbb{D} \setminus \{0\}$ and let ∇ be a flat, holomorphic connection on $\mathcal{V}_{|\mathbb{D}^*}$, such that* 0 *is a regular point of ∇. Then we can associate to ∇ an element*

$$\operatorname{Res}_0(\nabla) \in \operatorname{End}(\mathcal{V}_0)$$

in a well-defined way. Namely, let γ be a generator of $\pi_1(\mathbb{D}^*, x)$ (for some fixed $x \in \mathbb{D}^*$), let $f_1, \ldots, f_r$ be a local basis of $\mathcal{V}$ about 0*, and let*

$\Gamma = (\omega_{ij})_{i,j=1,\ldots,r}$ *be the matrix of* ∇ *associated with this basis (where* $\omega_i j$ *are sections of* $\Omega^1_{\overline{\mathcal{C}}}(S)$*). Then* $\mathrm{Res}_0(\nabla)$ *with respect to the basis* $\{f_i\}_{i=1}^r$ *is given by the matrix*

$$\mathrm{Res}_0(\Gamma) = \begin{pmatrix} \int_\gamma \omega_{11} & \cdots & \int_\gamma \omega_{1r} \\ \vdots & & \vdots \\ \int_\gamma \omega_{r1} & \cdots & \int_\gamma \omega_{rr} \end{pmatrix}.$$

PROOF: Let $g_1, \ldots, g_r$ be another basis about 0. We have to check that the matrix Γ' of ∇ with respect to $\{g_i\}_{i=1}^r$ gives the same endomorphism. Let A be the base change from $\{f_i\}$ to $\{g_i\}$. Then by Remark 6.11,

$$\Gamma' = A^{-1}\mathrm{d}A + A^{-1}\Gamma A,$$

hence the principal part of Γ' is conjugate by A^{-1} to the principal part of Γ. Therefore $A^{-1}\,\mathrm{Res}_0(\Gamma)A = \mathrm{Res}_0(\Gamma')$. ∎

Proposition 7.24 ([Del70, Theorem 1.17])
In the situation of Lemma 7.23, the monodromy transformation T *in* $\mathrm{Aut}(\mathcal{V}_{|\mathbb{D}^*})$ *associated with* $\gamma \in \pi_1(\mathbb{D}^*, x)$ *extends to an automorphism* $\tilde{T}$ *of* $\mathcal{V}$*, whose stalk at* 0 *is given by*

$$\tilde{T}_0 = \exp(-2\pi i\,\mathrm{Res}_0(\Gamma)).$$

Under some mild assumptions, we can extend a vector bundle associated with a local system on the pre-compact Riemann surface $\mathcal{C}$ over the cusps of $\mathcal{C}$. It suffices to apply the following proposition locally to a neighborhood of each cusp. The resulting extension of the vector bundle is called *Deligne extension.*

Proposition 7.25 ([Del70, Proposition 5.2])
Let $\mathbb{V}$ *be a* $\mathbb{C}$*-local system on* $\mathbb{D}^* = \mathbb{D}\setminus\{0\}$*, which is supposed to have unipotent monodromy about* 0*. Let* $\mathcal{V} = \mathbb{V} \otimes_{\mathbb{C}} \mathcal{O}_{\mathbb{D}^*}$*. Then there is a unique extension of* $\mathcal{V}$ *to a vector bundle* $\tilde{\mathcal{V}}$ *to* $\mathbb{D}$ *such that*

a) the matrix Γ *of the connection (with respect to any local basis of* $\tilde{\mathcal{V}}$*) has at most poles of order* 1 *at* 0*,*

b) the residue $\mathrm{Res}_0(\Gamma)$ *is nilpotent*

Kodaira-Spencer Map

We describe the Kodaira-Spencer map for a pure, polarized VHS of weight 1. Note that this map is the Higgs field in the language of Higgs bundles; a fact, which will not be discussed here (see [VZ04], [Möl06]).

Remark 7.26 Let V be a $2g$-dimensional $\mathbb{R}$-vector space, and let $W \subset V \otimes_{\mathbb{R}} \mathbb{C}$ be a $\mathbb{C}$-subvector space such that $V \otimes_{\mathbb{R}} \mathbb{C} = W \oplus \overline{W}$. Assume that we are given a $\mathbb{C}$-bilinear alternating form Q on V, such that $Q(W,W) = 0$ and $iQ(w,\overline{w}) > 0$ for all $w \in W \setminus \{0\}$. Then $V/W \to W^*$, $v + W \mapsto Q(\cdot, v)$ is a $\mathbb{C}$-linear isomorphism. We can sheafify this statement: If $\mathcal{V}^{1,0} \subset \mathcal{V}$ is the $(1,0)$-part of a VHS, polarized by Q, then

$$\mathcal{V} \,/\, \mathcal{V}^{1,0} \cong (\mathcal{V}^{1,0})^*.$$

Definition 7.27 Let $(\mathbb{V}, \mathcal{V}^{1,0}, Q)$ be a pure, polarized VHS of weight 1 on $\mathcal{C}$ with associated connection ∇. By Proposition 7.25, we can extend $\mathbb{V} \otimes \mathcal{O}_{\mathcal{C}}$ to a vector bundle $\mathcal{V}_{\text{ext}}$ on $\overline{\mathcal{C}}$, thus a fortiori, $\mathcal{V}^{1,0}$ has an extension $\mathcal{V}^{1,0}_{\text{ext}}$.

The Kodaira-Spencer map $\overline{\nabla}$ is the $\mathcal{O}_{\overline{\mathcal{C}}}$-linear map defined as the composition

$$\begin{aligned} \mathcal{V}^{1,0}_{\text{ext}} &\hookrightarrow \mathcal{V}_{\text{ext}} \xrightarrow{\nabla} \mathcal{V}_{\text{ext}} \otimes_{\mathcal{O}_{\overline{\mathcal{C}}}} \Omega^1_{\overline{\mathcal{C}}}(S) \longrightarrow \\ &\longrightarrow \mathcal{V}_{\text{ext}} \,/\, \mathcal{V}^{1,0}_{\text{ext}} \otimes_{\mathcal{O}_{\overline{\mathcal{C}}}} \Omega^1_{\overline{\mathcal{C}}}(S) \longrightarrow (\mathcal{V}^{1,0}_{\text{ext}})^* \otimes_{\mathcal{O}_{\overline{\mathcal{C}}}} \Omega^1_{\overline{\mathcal{C}}}(S). \end{aligned}$$

7.4 Characterization of Teichmüller Curves by their VHS

We summarize M. Möller's results on the VHS of a Teichmüller curve. Starting from P. Deligne's semisimplicity result, he deduces the following description.

Proposition 7.28 ([Möl06, Proposition 2.4])
Let (X,ω) be a Veech surface, and assume that $\Gamma \leqslant \Gamma(X,\omega)$ fulfills Condition $(*)$ from Remark 5.4. Let $L \subset \mathbb{C}$ be a Galois closure of the trace

field $K(X,\omega)$*, and* $r = (K(X,\omega) : \mathbb{Q})$*. The local system* $\mathbb{V} = R^1\phi_*(\mathbb{Z})$ *associated with the family* $\phi : \mathcal{X} \to \mathcal{C} = \mathbb{H}/\Gamma$ *splits over* $\mathbb{Q}$ *as*

$$\mathbb{V}_{\mathbb{Q}} = \mathbb{W}_{\mathbb{Q}} \oplus \mathbb{M}_{\mathbb{Q}}\ , \quad \textit{with} \quad \mathbb{W}_L = \mathbb{L}_1 \oplus \cdots \oplus \mathbb{L}_r\,.$$

Each $\mathbb{L}_i$ *carries a polarized* L*-VHS of weight* 1*, and* $\mathbb{M}_{\mathbb{Q}}$ *carries a polarized* $\mathbb{Q}$*-VHS of weight* 1*. Their sum is the pVHS on* $\mathbb{V}$*. Moreover, none of the* $(\mathbb{L}_i)_{\mathbb{C}}$ *is contained in* $\mathbb{M}_{\mathbb{C}}$*.*

He also derives from this result a characterizations of the image of a Teichmüller curve in $\mathcal{A}_g$. It is contained in the locus of abelian varieties that admit a splitting $A_1 \times A_2$ up to isogeny, where A_1 has real multiplication with the trace field $K(X,\omega)$ [Möl06, Theorem 2.7]. Since the trace field of an origami curve is $\mathbb{Q}$, and the first factor of the splitting comes from the covering map to the elliptic curve, this characterizations has no implications for origami curves.

More important for our purposes are the following two characterizations of Teichmüller curves.

Proposition 7.29 ([Möl06, Theorem 2.13])
Let $\Gamma \leqslant \mathrm{PSL}_2(\mathbb{R})$ *be a cofinite Fuchsian group, and let* $\phi : \mathcal{X} \to \mathcal{C} = \mathbb{H}/\Gamma$ *be a family of curves of genus* g*. Suppose that the local system* $R^1\phi_*(\mathbb{R}_{\mathcal{X}})$ *admits a direct summand* $\mathbb{L}$ *of rank 2, whose monodromy representation* ρ *is (up to conjugation in* $\mathrm{PSL}_2(\mathbb{R})$*) the Fuchsian embedding. Then* $\mathcal{C} \to \mathcal{M}_g$ *is a finite cover of a Teichmüller curve.*

Proposition 7.30 ([Möl06, Theorem 5.3])
Let Γ *be a cofinite Fuchsian group and let* $\phi : \mathcal{X} \to \mathcal{C} = \mathbb{H}/\Gamma$ *be a family of curves of genus* g *such that* $\mathbb{V} = R^1\phi_*\mathbb{Z}_{\mathcal{X}}$ *has unipotent monodromy about the cusps. Suppose* $\mathbb{V} \otimes_{\mathbb{Z}} \mathbb{C}$ *has a rank 2-subsystem* $\mathbb{L}$ *carrying a polarized VHS, whose Kodaira-Spencer map* $\overline{\nabla}$ *is an isomorphism. Then* $\mathcal{C} \to \mathcal{M}_g$ *is a finite cover of a Teichmüller curve.*

These two statements were partly reproved by A. Wright [Wri11, Theorem A.1, A.3] in a more down-to-earth manner. Note that the second statements is better formulated in terms of Higgs bundles, and that it gives an algebraic characterisation of Teichmüller curves. Note also that the assertion that $\overline{\nabla}$ be an isomorphism depends on the set of cusps.

Using Proposition 7.30, I. Bouw and M. Möller [BM10b] were able to find new Teichmüller curves, whose Veech groups are essentially all $\Delta(n, m, \infty)$-triangle groups. This family of Teichmüller curves extends Veech's and Ward's family [Vee89] and substantially overlaps with a family later described explicitly by P. Hooper [Hoo09] by giving the associated translation surfaces. They also reappear in [Wri11].

8 Multiplicative Ergodic Theory

8.1 Oseledets' Theorem

We recall some basic notions in ergodic theory, and in particular state the multiplicative ergodic theorem, first proved by Oseledets. A general reference for this part is [Wal82].

Recall that a *measure space* is a triple (X, Σ, m) of a set X, a σ-algebra Σ and a countably additive measure m. In the following, m will mostly be a finite measure, i. e. $m(X) < \infty$ and usually normalized to $m(X) = 1$ (in which case (X, Σ, m) is called a *probability space*). We will always assume that m is *complete*, i. e. $m(A) = 0$ implies $B \in \Sigma$ for all $B \subset A$. A *measurable function* $f : (X, \Sigma, m) \to (X', \Sigma', m')$ satisfies $f^{-1}(B) \in \Sigma$ for all $B \in \Sigma'$. A topological space will usually be equipped with its Borel σ-algebra, generated by the open sets. The space of integrable measurable functions $f : X \to \mathbb{R}$ will be denoted $L^1(X, m)$.

A *measure-preserving transformation* is a measurable $\mathbb{Z}$- or $\mathbb{R}$-action T_t on (X, Σ, m) such that $m(T_t(A)) = m(A)$ for all t and all $A \in \Sigma$ (endow $\mathbb{Z}$ with the counting measure and $\mathbb{R}$ with the Lebesgue-measure). A measurable $\mathbb{R}$-action is usually called *flow*. A measure m is called *ergodic* w. r. t. to a measure-preserving transformation T_t if every T_t-invariant $A \in \Sigma$ has $m(A) = 1$ or $m(A) = 0$. Ergodicity of m is equivalent to every T_t-invariant function being constant m-a.e.

Definition 8.1 Let (X, Σ, m) be a measure space and let V be an r-dimensional $\mathbb{R}$-vector space. Let g_t be a flow on X. A *cocyle* for the flow g_t is a measurable map

$$A : \mathbb{R} \times X \to \mathrm{GL}(V) \qquad (t, x) \mapsto A_t(x)$$

such that

$$A_{s+t}(x) = A_s(g_t(x)) \circ A_t(x)$$

holds for all s, $t \in \mathbb{R}$ and $x \in X$.

Theorem 8.2 (Oseledets) *Let (X, Σ, m) be a measure space with finite measure, g_t a measure-preserving flow, V an r-dimensional $\mathbb{R}$-vector space, endowed with a norm $\|\cdot\|$ and $A : \mathbb{R} \times X \to \mathrm{GL}(V)$ a cocycle for g_t such that*

$$\sup_{-1 \leq u \leq 1} \log^+(\|A_u(x)\|) \in L^1(X, m).$$

Then there is a measurable subset $U \subset X$ of full measure, invariant under g_t such that for every $x \in U$, there is $s(x) \in \mathbb{N}$ and real numbers

$$\lambda_1(x) > \lambda_2(x) > \ldots > \lambda_{s(x)}(x)$$

and a decomposition $V = \bigoplus_{i=1}^{s(x)} W_{i,x}$ such that for $1 \leq i \leq s(x)$ and $v \in W_{i,x} \setminus \{0\}$

$$\lim_{t \to \infty} \frac{1}{t} \log \|A_t(x) \cdot v\| = \lambda_i(x).$$

Moreover, s and λ_i are (when defined) measurable and g_t-invariant and the $W_{i,x}$ are measurable subbundles of $U \times V$.

PROOF: Combine [Rue79, Theorem 3.1] with [Rue79, Theroem B.3]. ■

Definition 8.3 In the situation of Theorem 8.2, the numbers $\lambda_i(x)$, $i = 1, \ldots, s(x)$ are called the *Lyapunov exponents* for A_t and g_t at $x \in X$. The number $m_i(x) = \dim_{\mathbb{R}} W_{i,x}$ is called the *multiplicity* of $\lambda_i(x)$.

Remark 8.4 In the situation of Theorem 8.2, if the measure m is ergodic w. r. t. to the flow g_t, then s, λ_i and m_i are constant almost everywhere. In this case, we call the collection $(\lambda_i)_{i=1}^{\dim_{\mathbb{R}} V}$ (where each λ_i appears m_i times) the *Lyapunov spectrum* of A. If $m_i = 1$, then λ_i will be called *simple.*

8.2 The Kontsevich-Zorich Cocycle

The Kontsevich-Zorich cocycle is a cocycle over the Teichmüller flow on $\Omega\mathcal{M}_g$. In order to give its definition, we first need a finite measure on the base space. To this end, recall the definition of the measure ν from Section 2.2 defined on a stratum $\Omega\,\mathcal{T}_g(\kappa)'$ given by the partition κ of $2g-2$. Veech [Vee90] showed that the disintegration of ν w. r. t. the unit hyperboloid $\Omega_1\,\mathcal{T}_g(\kappa)'$ of norm 1 surfaces descends to a measure ν_1 on the quotient $\Omega_1\mathcal{M}_g(\kappa)$. This measure has the following properties:

Remark 8.5 a) The total ν_1-mass of any connected component in any stratum is finite.

b) ν_1 is an $\mathrm{SL}_2(\mathbb{R})$-invariant measure; it is ergodic for the actions of both $\mathrm{SL}_2(\mathbb{R})$ and the Teichmüller flow $\{g_t\}$ on every connected component of every stratum of $\Omega_1\,\mathcal{M}_g$.

These assertions were independently shown by H. Masur [Mas82] and W. Veech [Vee82], [Vee90]. We remark that A. Eskin and A. Okounkov [EO01] have computed the volumes of the strata in terms of quasi-modular forms by counting origamis in the strata.

Having constructed this measure, one is led to consider more generally any $\mathrm{SL}_2(\mathbb{R})$-invariant, finite measure on $\Omega_1\mathcal{M}_g(\kappa)$.

In our definition of the Kontsevich-Zorich cocycle, we follow Forni [For06]. In the following remark, let the mapping class group $\Gamma(S)$ act on cohomology in the usual way, i. e. f sends a cohomology class v to $(f^{-1})^*(v)$. Note also that we do not obtain a cocycle in the true sense of our definition, but rather a flow on a (trivial) vector bundle that is a lift of the flow on the base.

Remark 8.6 Let S be a compact Riemann surface of genus g with n marked points, such that $3g-3+n>0$. Let

$$\widehat{G^{\mathrm{KZ}}} : \mathbb{R}\times\Omega_1\,\mathcal{T}(S)\times H^1(S,\mathbb{R}) \to \Omega_1\,\mathcal{T}(S)\times H^1(S,\mathbb{R})\,, \quad (t,x,v)\mapsto (g_t(x),v)$$

be the trivial cocycle for the Teichmüller geodesic flow. Consider the orbifold vector bundle over $\Omega_1\,\mathcal{M}_{g,n}$

$$\mathcal{H}\,\mathcal{M}_{g,n} = (\Omega_1\,\mathcal{T}(S)\times H^1(S,\mathbb{R}))/\,\Gamma(S).$$

Then $\widehat{G^{\mathrm{KZ}}}$ descends to a quotient flow

$$G_t^{\mathrm{KZ}} : \mathcal{H}\,\mathcal{M}_{g,n} \to \mathcal{H}\,\mathcal{M}_{g,n}\,.$$

The vector bundle $h : \Omega_1 \mathcal{T}(S) \times H^1(S, \mathbb{R}) \to \Omega_1 \mathcal{T}(S)$ carries a norm. In the fiber over $x = (X, m : S \to X, \omega) \in \Omega_1 \mathcal{T}(S)$, a vector $v \in h^{-1}(x)$ has norm

$$\|v\|_m = \|(m^{-1})^*(v)\|_X ,$$

where $\| \cdot \|_X$ is the Hodge-norm on X, scaled by $\frac{1}{2}$ (see Remark 4.12). $\| \cdot \|_m$ is invariant under the action of $\Gamma(S)$ on $\Omega_1 \mathcal{T}(S) \times H^1(S, \mathbb{R})$.

Proof: Let f be a mapping class, and let $(x, v) \in \Omega_1 \mathcal{T}(S) \times H^1(S, \mathbb{R})$, $x = (X, m, \omega)$. Then $g_t(x) = (g_t \cdot X, \varphi_{g_t} \circ m, g_t \cdot \omega)$ and $f(x) = (X, m \circ f^{-1}, \omega)$, so $f(g_t(x)) = g_t(f(x))$. Moreover, for a vector $v \in h^{-1}(x)$,

$$\|(f^{-1})^* v\|_{m \circ f^{-1}} = \|v\|_m.$$

■

Definition 8.7 G_t^{KZ} is called the *Kontsevich-Zorich cocycle* for the Teichmüller geodesic flow.

While it may appear that $\widehat{G_t^{\mathrm{KZ}}}$ is trivial, this is not really the case, since the norm is not equivariant for the Teichmüller flow. Note also that $\Omega_1 \mathcal{T}(S) \times H^1(S, \mathbb{R})$ is a trivialization of the (pullback of the) $\mathbb{R}$-local system $R^1 \phi_{\mathrm{univ}*}(\mathbb{R}_{\mathcal{X}_{\mathrm{univ}}})$ on $\mathcal{T}(S)$ induced by the universal family $\phi_{\mathrm{univ}} : \mathcal{X}_{\mathrm{univ}} \to \mathcal{T}(S)$ over Teichmüller space.

Remark 8.8 For simplicity, we will further restrict to $n = 0$ and $g \geq 2$. Let μ be a g_t-invariant, finite, ergodic measure on $\Omega_1 \mathcal{M}_g$. We would like to apply Theorem 8.2 to G_t^{KZ}. However, the vector bundle $\Omega_1 \mathcal{T}(S) \times H^1(S, \mathbb{R})$ is *not* trivial as a *normed* vector bundle; in particular, a priori the cocycle does not live on a fixed normed vector space. On the other hand, every measurable normed vector bundle over a compact metric space can be trivialized on a set of full measure. Unfortunately, the base is a non-compact metric space, even after passing to the quotient by the mapping class group. Nevertheless, things can be made to work, using the analysis of W. Veech [Vee86] and A. Zorich (see [Zor96] and his survey [Zor06] for a nice presentation) of the Teichmüller flow by means of zippered rectangles and Rauzy-Veech induction.

The next remark assembles fundamental results on the Lyapunov spectrum of G_t^{KZ}.

Remark 8.9 Let $g \geq 2$, and let μ be a g_t-invariant, finite, ergodic measure on $\Omega_1 \mathcal{M}_g$. Theorem 8.2 and Remark 8.6 allow us to speak of the Lyapunov exponents of G_t^{KZ} with respect to μ. The Lyapunov spectrum

$$\lambda_1^\mu \geq \lambda_2^\mu \geq \ldots \geq \lambda_{2g}^\mu$$

of G_t^{KZ} has the following properties

a) The spectrum is symmetric with respect to 0, i. e. $\lambda_{g+k}^\mu = -\lambda_{g-k+1}^\mu$ for all $k = 1, \ldots, g$. Therefore, we will henceforth speak of

$$\lambda_1^\mu \geq \ldots \geq \lambda_g^\mu \geq 0$$

as of the *Lyapunov spectrum* of G_t^{KZ} w. r. t. to μ. In particular, when speaking of the non-negative Lyapunov exponents, we mean precisely the exponents contained in the Lyapunov spectrum.

b) The first exponent is always simple and equal to 1; it is therefore called *trivial Lyapunov exponent.*

c) The spectrum for the measure ν_1 is simple and non-degenerate, i. e.

$$1 = \lambda_1^{\nu_1} > \lambda_2^{\nu_1} > \ldots > \lambda_g^{\nu_1} > 0.$$

Part a) follows from the fact that the action of $\Gamma(S)$ is by symplectic matrices on $H^1(S, \mathbb{R})$. We assemble some references for the remaining statements. Part b) was originally shown by W. Veech [Vee86] for the measure ν_1, and has been proved by G. Forni (e. g. [For06, Theorem 5.1]) for every ergodic probability measure on $\Omega_1 \mathcal{M}_g$. Part c) is the Zorich conjecture, formulated in [Zor96]. The conjecture was proved in several steps by different authors: Simplicity of the spectrum for ν_1 was proved in full generality by A. Avila and M. Viana [AV07]. Non-degenerateness (non-uniform hyperbolicity) was shown by Forni [For02].

We reformulate the Kontsevich-Zorich cocycle in the case of a Teichmüller curve. Let (X, ω) be a Veech surface of genus g, renormalized such that $\mathrm{Area}(X, \omega) = 1$, and consider its $\mathrm{SL}_2(\mathbb{R})$-orbit $M = \mathrm{SL}_2(\mathbb{R}) \cdot (X, \omega) \subset \Omega_1 \mathcal{M}_g$. Since M is closed, we obtain a finite measure μ_M on $\Omega_1 \mathcal{M}_g$ with support M, namely the measure induced from the Haar measure λ on $\mathrm{SL}_2(\mathbb{R})$. λ is $\mathrm{SL}_2(\mathbb{R})$-invariant and ergodic for g_t (see [CFS82, Chap.4,

§4, Theorem 1]), and μ_M inherits these two properties. Consider a Teichmüller disk Δ associated with (X, ω), e. g. the one defined by $(X, \mathrm{id}, \omega^{\otimes 2})$ in $\mathcal{T}(X)$. Let $U\Delta = \mathrm{SL}_2(\mathbb{R}) \cdot (X, \mathrm{id}, \omega) \subset \Omega_1 \mathcal{T}(X)$ be the "unit tangent bundle" to Δ.

Definition 8.10 In the situation above, the *Kontsevich-Zorich cocycle* for the Teichmüller curve associated with (X, ω) is defined as the quotient cocycle $G_t^{\mathrm{KZ}}(X, \omega)$ by the action of $\mathrm{Aff}(X, \omega)$ of the cocycle

$$\mathbb{R} \times U\Delta \times H^1(X, \mathbb{R}) \to U\Delta \times H^1(X, \mathbb{R}) \;, (t, x, v) \mapsto (g_t(x), v)$$

on the orbifold vector bundle $\mathcal{H} M = (U\Delta \times H^1(X, \mathbb{R}))/\,\mathrm{Aff}(X, \omega)$ over $M = \mathrm{SL}_2(\mathbb{R}) \cdot (X, \omega) \subset \Omega_1 \mathcal{M}_g$.

This cocycle is morally a restriction of the big cocycle on $\Omega_1 \mathcal{M}_g$ with the restriction that we deal with orbifold vector bundles for different groups, so one has to be careful about possible identifications.

Remark 8.11 Let $(X, \omega) \in \Omega_1 \mathcal{M}_g$ be a Veech surface and $M = \mathrm{SL}_2(\mathbb{R}) \cdot (X, \omega)$.

a) Since the flow g_t is ergodic on M, the Lyapunov exponents are well-defined. The *Lyapunov spectrum* is the collection of the g real numbers

$$\lambda_1^{\mu_M} = 1 > \lambda_2^{\mu_M} \geq \ldots \geq \lambda_g^{\mu_M}.$$

b) The Lyapunov spectrum of $G_t^{\mathrm{KZ}}(X, \omega)$ does not change, if we consider the induced cocycle on a quotient $(U\Delta \times H^1(X, \mathbb{R}))/\Gamma$ by a finite-index subgroup $\Gamma \leqslant \mathrm{Aff}(X, \omega)$.

c) Let $\Gamma \leqslant \mathrm{Aff}(X, \omega)$ be a subgroup of finite index. To any symplectic Γ-invariant subspace $V \subset H^1(X, \mathbb{R})$ of dimension $2r$ are associated r of the Lyapunov exponents of $G_t^{\mathrm{KZ}}(X, \omega)$ in a)

$$\lambda_{i(1)} \geq \cdots \geq \lambda_{i(r)}.$$

PROOF: b) follows from the fact that we deal with a quotient cocycle. For Part c), note that by Γ-invariance of V, we obtain a subspace of a finite cover of the orbifold vector bundle $\mathcal{H} M$, where we can apply Oseledets

Theorem (modulo Remark 8.8). The resulting Lyapunov exponents will again be symmetric w. r. t. 0, because of the symplectic structure of V, and the Lyapunov spectrum (i. e. the "non-negative" part) will be a subset of the whole spectrum by b). ■

Remark 8.12 Following [EKZ10a], we explain another informal approach to the Lyapunov exponents of the Kontsevich-Zorich cocycle of a Teichmüller curve.

Let (X, ω) be a Veech surface. Consider a "generic" geodesic c of the flow on the Teichmüller curve $\mathbb{H} / \operatorname{Aff}(X, \omega)$, i. e. one that winds "ergodically", so in particular, minimally around the surface, and does not run into a cusp. At every time t that our geodesic comes close to the initial point, we can close it up artificially and obtain a closed geodesic, i. e. an element γ of $\operatorname{Aff}(X, \omega)$. Let $A(t) = \rho(\gamma)$ be the associated monodromy matrix. Then the limit

$$\lim_{t \to \infty} (A(t)^T A(t))^{1/(2t)} = \Lambda$$

exists and the Lyapunov exponents of the Kontsevich-Zorich cocycle are given as the logarithms of eigenvalues of the matrix Λ. In particular, since $A(t)$ is a symplectic matrix, the Lyapunov exponents are symmetric w. r. t. 0.

8.3 The Lyapunov Spectrum of Teichmüller Curves

There is an algebraic formula for the sum of the non-negative Lyapunov exponents of G_t^{KZ} of a rank 2-subbundle or more generally on the determinant bundle of a higher rank-bundle: the formula for the sum of the Lyapunov exponents. It was first discovered by Kontsevich and Zorich [Kon97], and has been formulated in various forms by different authors (compare also with Proposition 9.14). We present a result due to Eskin, Kontsevich and Zorich [EKZ10b] which is adapted to origamis.

Proposition 8.13 (Sum of the Lyapunov Exponents for origamis) *Let (X, ω) be an origami of genus g in the stratum $\Omega\mathcal{M}(\kappa)$. Define*

$$c(\kappa) = \frac{1}{12} \sum_{i=1}^{\ell(\kappa)} \frac{\kappa_i(\kappa_i + 2)}{\kappa_i + 1},$$

and

$$c(X,\omega) = (\mathrm{SL}_2(\mathbb{Z}) : \Gamma(X,\omega))^{-1} \cdot \sum_{Y \in \mathrm{SL}_2(\mathbb{Z})\cdot(X,\omega)} \Big(\sum_{i \in I_Y} \frac{h_{C_{Y,i}}}{w_{C_{Y,i}}} \Big),$$

where the innermost sum runs over a decomposition of the surface Y into horizontal cylinders $\{C_{Y,i}\}_{i\in I_Y}$ of height $h_{C_{Y,i}}$ and width $w_{C_{Y,i}}$. The Lyapunov spectrum $\{\lambda_i\}_{i=1}^{g}$ of $G_t^{\mathrm{KZ}}(X,\omega)$ obeys to the sum formula

$$\sum_{i=1}^{g} \lambda_i = c(\kappa) + c(X,\omega).$$

Remark 8.14 On some strata in low genus, the sum of the Lyapunov exponents is non-varying, i. e. constant for all Teichmüller curves in that stratum. This phenomenon was first discovered with the help of computer experiments in [Kon97] and has recently been proved by D. Chen and M. Möller [CM11].

Proposition 8.15 ([Bai07, Theorem 1.7])
For any $\mathrm{SL}_2(\mathbb{R})$-invariant ergodic measure μ on $\Omega_1\mathcal{M}_2$, the non-trivial Lyapunov exponent is

- *equal to $\frac{1}{2}$ if the measure is supported on $\Omega\mathcal{M}_2(1,1)$, and is*
- *equal to $\frac{1}{3}$ if the measure is supported on $\Omega\mathcal{M}_2(2)$.*

This is an unpublished result by M. Kontsevich and A. Zorich; it was also proved for the case of Teichmüller curves by I. Bouw and M. Möller [BM10a, Corollary 2.4].

There are several examples of Teichmüller curves, where one can explicitly give the Lyapunov spectrum. In [BM10b], I. Bouw and M. Möller find algebraically primitive Teichmüller curves, most of them unknown before, and determine the associated Lyapunov spectrum. In [EKZ10a], we are given the Lyapunov spectra of cyclic pillowcase covers; the same is done for abelian pillow-case covers in [Wri11]. There are very particular Teichmüller curves, one in $\mathcal{M}_3$, one in $\mathcal{M}_4$, where all of the non-trivial Lyapunov exponents vanish. The curve in $\mathcal{M}_3$, also called "Eierlegende Wollmilchsau" was first investigated by Herrlich and Schmithüsen [HS08].

M. Möller [Möl05a] proved that both curves are the only examples of Shimura-Teichmüller curves (with the possible exception of curves in genus 5); their Lyapunov spectrum is examined [For02], and also in [FMZ10].

Let us mention one particular result in [EKZ10a], since it regards the Lyapunov spectrum of the stairs origamis $\mathbf{St}_n$ (see Section 3.2).

Proposition 8.16 ([EKZ10a, Proposition 2])
The Lyapunov spectrum of $G_t^{KZ}(\mathbf{St}_n)$ for the stairs origami $\mathbf{St}_n$ is

$$\frac{1}{n}, \frac{3}{n}, \frac{5}{n}, \ldots, \frac{n}{n}, \qquad \text{if } n \text{ is odd}$$
$$\frac{2}{n}, \frac{4}{n}, \frac{6}{n}, \ldots, \frac{n}{n}, \qquad \text{if } n \text{ is even}$$

As stated above, the Lyapunov spectrum of a Teichmüller curve can partly vanish. We give a necessary condition for this situation. M. Möller has communicated to the author that it is not a sufficient condition.

Proposition 8.17 *Let (X, ω) be a Veech surface of genus g, and let W be an r-dimensional subspace, invariant under the action of a finite-index subgroup Γ of $\mathrm{Aff}(X, \omega)$.*

a) If Γ acts on W by unitary matrices for the Hodge inner product H, then Γ acts as a finite group.

b) If Γ acts on W as a finite group, then all of the Lyapunov exponents associated with W vanish.

PROOF: a) The action of $\mathrm{Aff}(X, \omega)$ on $H^1(X, \mathbb{R})$ is discrete. Hence Γ acts on W by a discrete subgroup of the unitary group. The latter being compact implies that Γ acts by a finite group.

b) Passing to a finite cover, we can achieve that Γ acts trivially on W. This implies the statement.

Remark 8.18 Let (X, ω) be a Veech surface, and let $\phi : \mathcal{X} \to \mathcal{C} = \mathbb{H}/\Gamma$ be the family over the Teichmüller curve associated with (X, ω), with $\Gamma \leqslant \mathrm{Aff}(X, \omega)$ fulfilling Condition $(*)$. Let W be the maximal subspace of $H^1(X, \mathbb{Q})$ on which Γ acts trivially. Then the local system $\mathbb{W} \subset R^1\phi_*(\mathbb{Q})$

associated with W is the constant sheaf of fiber W. By [Del71a, Corollaire 4.1.2], $\mathbb{W}$ carries a sub-pVHS of $(R^1\phi_*(\mathbb{Q}), \phi_*\Omega^1_{\mathcal{X}/\mathcal{C}}, Q)$, and the induced Hodge structure on $\mathbb{W}_c$ is independent of $c \in \mathcal{C}$. Thus it leads to a fixed part in the family of Jacobians over the Teichmüller curve, i. e. an inclusion $A \times \mathcal{C} \to \mathrm{Jac}(\mathcal{X} \to \mathcal{C})$ with a fixed abelian variety A (see also [Bau09]).

9 Splitting the Hodge Bundle over a Teichmüller curve

From Theorem 7.15 we know that the pVHS on a compact complex manifold B is completely reducible. There is a remarkable theorem by Deligne which states that for a fixed base, only finitely many different monodromy representations can occur.

Theorem 9.1 ([Del87, Théorème 0.1])
Let B be a fixed smooth, connected algebraic variety over $\mathbb{C}$, let $b \in B$, and $n \in \mathbb{N}$. Then for $k \in \mathbb{N}_0$ variable, the monodromy representations $\rho : \pi_1(B,b) \to H^k(\mathcal{X}_b, \mathbb{Q})$ of dimension n coming from algebraic families $\mathcal{X} \to B$ fall into finitely many isomorphism classes.

Deligne actually proves that for a fixed base B and $n \in \mathbb{N}$, the local systems of $\mathbb{Q}$-vector spaces of dimension n, which are direct summands of a local system associated with a pVHS, fall into finitely many isomorphism classes.

However, not much is known about the representation, respectively local systems that actually do occur – even for families of curves, and nor do we to my knowledge know how to obtain all of them.

In this chapter, we first present two basic concepts to obtain subvector spaces of the first cohomology that are invariant under the action of (a finite-index subgroup of) $\mathrm{Aff}(X,\omega)$ and therefore permit to decompose the monodromy representation (respectively the local system). One is the use of translation coverings, the second is the use of representation theory for the finite group $\mathrm{Aut}(X,\omega)$. After this, we present a method for computing the Lyapunov exponent of a rank 2-subrepresentation, based on an outline of M. Möller.

Finally, we apply the two concepts to origamis. This part relies on computations carried out with the help of the origami program, which was developed at our workgroup mainly by G. Weitze-Schmithüsen, K. Kremer, M. Finster and myself.

9.1 Coverings

Let $p : (X,\omega) \to (Y,\nu)$ be a Veech covering between Veech surfaces. By [GJ00, Theorem 4.8] the elements of $\mathrm{Aff}(X,\omega)$ that descend via p to Y form a finite-index subgroup $\mathrm{Aff}(X,\omega)_p$ of $\mathrm{Aff}(X,\omega)$. Let

$$\varphi_p : \mathrm{Aff}(X,\omega)_p \to \mathrm{Aff}(Y,\nu)$$

be the group homomorphism that maps $f \in \mathrm{Aff}(X,\omega)_p$ to $\overline{f} \in \mathrm{Aff}(Y,\nu)$ such that $p \circ f = \overline{f} \circ p$. The image of φ_p is the finite-index subgroup $\mathrm{Aff}(Y,\nu)^p$ of $\mathrm{Aff}(Y,\nu)$ of affine diffeomorphisms, that lift to (X,ω).

Proposition 9.2 *Let $p : (X,\omega) \to (Y,\nu)$ be a Veech covering between Veech surfaces. Let*

$$\rho : \mathrm{Aff}(X,\omega) \to \mathrm{Sp}(H^1(X,\mathbb{Q})) \ , \quad f \mapsto (f^{-1})^*$$

be the monodromy representation of (X,ω). Then the image U of $H^1(Y,\mathbb{Q})$ under

$$p^* : H^1(Y,\mathbb{Q}) \to H^1(X,\mathbb{Q})$$

is an $\mathrm{Aff}(X,\omega)_p$-invariant symplectic subspace of $H^1(X,\mathbb{Q})$.

The map p^* is equivariant for the action of $\mathrm{Aff}(X,\omega)_p$ on U and $\mathrm{Aff}(Y,\nu)^p = \varphi_p(\mathrm{Aff}(X,\omega)_p)$ on $H^1(Y,\mathbb{Q})$.

Proof: Let $f \in \mathrm{Aff}(X,\omega)_p$ and $\overline{f} \in \mathrm{Aff}(Y,\nu)$ such that $p \circ f = \overline{f} \circ p$. Then for every $c \in H^1(Y,\mathbb{Q})$

$$(f^{-1})^*(p^*(c)) = (p \circ f^{-1})^*(c) = (\overline{f}^{-1} \circ p)^*(c) = p^*((\overline{f}^{-1})^*(c)) \ ,$$

proving $(f^{-1})^*(\mathrm{Im}(p^*)) \subset \mathrm{Im}(p^*)$. The computation also shows that p^* is equivariant. Finally, p^* is a symplectic map (see Example 4.2). ■

Theorem 9.3 *Let (X,ω), (Y,ν) be Veech surfaces of genus $g = g(X)$ and $g' = g(Y)$, and let $p : (X,\omega) \to (Y,\nu)$ be a Veech covering. Then*

- *there is a finite-index subgroup* $\Gamma \leqslant \Gamma(X,\omega) \cap \Gamma(Y,\nu)$ *such that Condition (∗) is simultaneously verified for the covers*

$$j_X : \mathcal{C} \to \mathbb{H} / \operatorname{Aff}(X,\omega) \to \mathcal{M}_g \quad \textit{and} \quad j_Y : \mathcal{C} \to \mathbb{H} / \operatorname{Aff}(Y,\nu) \to \mathcal{M}_{g'}$$

  with $\mathcal{C} = \mathbb{H} / \Gamma$.

- *if* $\phi_X : \mathcal{X} \to \mathcal{C}$ *and* $\phi_Y : \mathcal{Y} \to \mathcal{C}$ *are the respective families of curves, there is an inclusion* $\Phi : R^1 {\phi_Y}_* \mathbb{Z} \to R^1 {\phi_X}_* \mathbb{Z}$ *of local systems, which is a morphism of pVHS.*

PROOF: To prove the first claim, let $\Gamma' \leqslant \operatorname{Aff}(Y,\nu)$ be a finite-index subgroup fulfilling Condition (∗) for (Y,ν). Since it is torsion-free, it maps isomorphically onto a finite-index subgroup $\Gamma'' \leqslant \Gamma(Y,\nu)$. Similarly, we find a finite-index subgroup Δ' fulfilling Condition (∗) for (X,ω) such that Δ' maps isomorphically onto a finite-index subgroup Δ'' of $\Gamma(X,\omega)$. By [GJ00, Theorem 4.9] $\Gamma(X,\omega)$ and $\Gamma(Y,\nu)$ are commensurate, so Γ'' and Δ'' are also commensurate. So take $\Gamma = \Gamma'' \cap \Delta''$.

For the second claim, we will use the following rigidity of pVHS [PS03, Corollary 12]: Let B be a complex manifold, embeddable in a compact complex manifold $\overline{B}$ such that $\overline{B} \setminus B$ is a divisor with normal crossings. Let $\mathbb{V}$, $\mathbb{W}$ be local systems on B carrying each a pVHS and let $b \in B$. Then any $\pi_1(B,b)$-equivariant morphism of Hodge structures $\Phi_b : \mathbb{V}_b \to \mathbb{W}_b$ extends to a morphism of pVHS $\Phi : \mathbb{V} \to \mathbb{W}$.

We apply this to $\mathcal{C} = \mathbb{H} / \Gamma$. Let $c \in \mathcal{C}$ be a point with $j_X(c) = [X] \in \mathcal{M}_g$ and $j_Y(c) = [Y] \in \mathcal{M}_{g'}$. Such a point exists, since we can choose particular Teichmüller embeddings $\tilde{\jmath}_X : \mathbb{H} \to \mathcal{T}(X)$ and $\tilde{\jmath}_Y : \mathbb{H} \to \mathcal{T}(Y)$ inducing j_X and j_Y that map $\tau = i \in \mathbb{H}$ to $(X, \mathrm{id}_X) \in \mathcal{T}(X)$, respectively to $(Y, \mathrm{id}_Y) \in \mathcal{T}(Y)$. Consider the map

$$\Phi_b : H^1(\mathcal{Y}_c, \mathbb{Z}) \cong H^1(Y,\mathbb{Z}) \to H^1(X,\mathbb{Z}) \cong H^1(\mathcal{X}_c, \mathbb{Z})$$

induced by p^*. Since p is holomorphic, the map p^* is a morphism of Hodge structures. It respects the polarizations if we let the one on the source be the original one multiplied with $\deg(p)$. Moreover, w. l. o. g. we may assume $\Gamma \leqslant D(\operatorname{Aff}(X,\omega)_p)$. It then follows from Proposition 9.2 and [Bau09, Lemma 2.4.3] that Φ_b is equivariant w. r. t. the two monodromy actions of Γ. Therefore, we obtain a morphism of pVHS Φ as desired. To see that it is an inclusion, we note that p^* and therefore Φ_b is a monomorphism in the category of π_1-representations, which by Proposition 6.4 is equivalent to the category of local systems. ■

9.2 Representations

Now we describe how one can find a splitting of $H^1(X, \mathbb{R})$ by means of representation theory of finite groups. A general reference for this subject is [Ser96]. In the following $\mathbb{K}[G]$ denotes the group ring of a group G with coefficients in the field $\mathbb{K}$.

Let X be a compact Riemann surface. Recall that $\mathrm{Aut}(X)$ is a finite group acting (from the left) on $\Omega^1_X(X)$. Let $G \leqslant \mathrm{Aut}(X)$ be a subgroup. We can write $\Omega^1_X(X)$ as a direct sum of $\mathbb{C}[G]$-isotypic components

$$\Omega^1_X(X) = \bigoplus_{\chi} V_\chi \ , \tag{9.1}$$

where χ runs through all irreducible characters of G. The G-action extends to $H^1(X, \mathbb{C}) = \Omega^1_X(X) \oplus \overline{\Omega^1_X(X)}$, and G preserves the splitting into holomorphic and antiholomorphic parts. Therefore, $H^1(X, \mathbb{C})$ decomposes as

$$H^1(X, \mathbb{C}) = \bigoplus_{\chi} V_\chi \oplus \bigoplus_{\chi} \overline{V_\chi}. \tag{9.2}$$

Note that we would have to group representations together to see the decomposition of $H^1(X, \mathbb{C})$ into isotypic components. Analogously, the real cohomology $H^1(X, \mathbb{R})$ is a completely reducible $\mathbb{R}[G]$-module, and we can write it as

$$H^1(X, \mathbb{R}) = \bigoplus_{\vartheta} W_\vartheta \ , \tag{9.3}$$

a direct sum of $\mathbb{R}[G]$-isotypic components. More generaly, such a decomposition of $H^1(X, \mathbb{K})$ exsits for any field $\mathbb{K}$ of characteristic 0.

In both cases, g acts on a cohomology class, respectively on a differential by $(g^{-1})^*$, which we will abbreviate by $g\cdot$. Note that the action of G on the cohomology is in fact defined over $\mathbb{Z}$.

The following, probably well-known lemma describes the relation between the two G-actions.

Lemma 9.4 *In the situation above,*

a) if $V \subset \Omega^1_X(X)$ is a $\mathbb{C}[G]$-submodule, then $V \oplus \overline{V}$ is defined over $\mathbb{R}$, i. e. there exists an $\mathbb{R}[G]$-submodule $W \subset H^1(X, \mathbb{R})$, such that $W \otimes_{\mathbb{R}} \mathbb{C} = V \oplus \overline{V}$.

b) if $W \subset H^1(X, \mathbb{R})$ is an $\mathbb{R}[G]$-isotypic component, then W is a sub-$\mathbb{R}$-Hodge structure of $H^1(X, \mathbb{R})$.

PROOF: a) We let W be the $\mathbb{R}$-vector space spanned by $v + \overline{v}$ with $v \in V$. Via the inclusion $H^1(X, \mathbb{R}) \to H^1(X, \mathbb{C})$, W is a subspace of $H^1(X, \mathbb{R})$, and a G-module, if we let $g \in G$ act by the rule $v + \overline{v} \mapsto g \cdot v + \overline{g \cdot v}$. We have $W \otimes_{\mathbb{R}} \mathbb{C} = V \oplus \overline{V}$, since $iv \in V$ and $i(iv + \overline{iv}) = \overline{v} - v$, and this implies

$$v = \frac{1}{2}(v + \overline{v} - i(iv + \overline{iv})) \in W \otimes_{\mathbb{R}} \mathbb{C} \ ,$$

and similarly $\overline{v} \in W \otimes_{\mathbb{R}} \mathbb{C}$. This proves the inclusion from the right to the left. The other one is trivial.

b) $W \otimes \mathbb{C}$ decomposes as a direct sum of (at most two) G-isotypic components of $H^1(X, \mathbb{C})$. The sum decomposition (9.2) shows that each of these isotypic components is the direct sum of a holomorphic and an antiholomorphic part. This proves that $W = W^{1,0} \oplus W^{0,1}$ with $W^{1,0} \subset \Omega^1_X(X)$ and $W^{0,1} = \overline{W^{1,0}}$. ■

Now we investigate the interplay between affine automorphisms and the affine group of a Veech surface.

Proposition 9.5 *Let (X, ω) be a Veech surface, let $G \leqslant \mathrm{Aut}(X, \omega)$, and let $\mathbb{K} = \mathbb{R}$ or $\mathbb{C}$. The action $\rho : \mathrm{Aff}(X, \omega) \to \mathrm{Sp}(H^1(X, \mathbb{K}))$, $f \mapsto (f^{-1})^*$, restricted to the normalizer $N(G)$ of G in $\mathrm{Aff}(X, \omega)$, permutes the G-isotypic components of the decomposition* (9.2), *respectively* (9.3), *and* *there is a finite index subgroup* $\Gamma \leqslant \mathrm{Aff}(X,\omega)$ *such that every isotypic component is* Γ*-invariant.*

PROOF: As $\mathrm{Aut}(X,\omega)$ is normal in $\mathrm{Aff}(X,\omega)$, the normalizer $N(G)$ of G in $\mathrm{Aff}(X,\omega)$ has finite index in $\mathrm{Aff}(X,\omega)$. For all $g \in G$, and $f \in N(G)$, there exists $\tilde{g} \in G$, such that $gf = f\tilde{g}$. Therefore for all irreducible $\mathbb{K}[G]$-submodules V of $H^1(X,\mathbb{K})$, we have

$$(g^*)^{-1} \circ (f^*)^{-1}(V) = ((gf)^*)^{-1}(V) = ((f\tilde{g})^*)^{-1}(V) = (f^*)^{-1}(V),$$

which shows that $(f^{-1})^*(V)$ is another irreducible $\mathbb{K}[G]$-module inside $H^1(X,\mathbb{K})$. Hence every $f \in N(G)$ induces a permutation of the isotypic components of the representation of G. Thus there is a finite index subgroup $\Gamma \leqslant N(G)$ that leaves every isotypic component invariant. ■

Note that we can also find a finite-index subgroup $\Gamma \leqslant \mathrm{Aff}(X,\omega)$ such that each element $\gamma \in \Gamma$ acts on each isotypic component V of the decomposition of $H^1(X,\mathbb{K})$ as a $\mathbb{K}[G]$-linear automorphism. Indeed, it suffices to take $\Gamma = \bigcap_{g\in G} C(g)$, where $C(g)$ is the centralizer of g in $\mathrm{Aff}(X,\omega)$.

Theorem 9.6 *Let* (X,ω) *be a Veech surface and let* $G \leqslant \mathrm{Aut}(X,\omega)$*. Suppose* $U \subset H^1(X,\mathbb{R})$ *is a* G*-isotypic component. Then there is a finite-index subgroup* $\Gamma \leqslant \mathrm{Aff}(X,\omega)$ *fulfilling Condition* $(*)$ *such that the pVHS* $R^1\phi_*(\mathbb{R})$ *associated with the family over the Teichmüller curve* $\phi : \mathcal{X} \to \mathcal{C} = \mathbb{H}/\Gamma$ *has an* $\mathbb{R}$*-local subsystem* $\mathbb{U}$ *induced by* U*, which carries a sub-*$\mathbb{R}$*-pVHS.*

PROOF: W. l. o. g. we can choose $\Gamma \leqslant \mathrm{Aff}(X,\omega)$ such that it fulfills Condition $(*)$ and leaves U invariant. Let $\mathbb{U}$ be the associated $\mathbb{R}$-local system on $\mathcal{C} = \mathbb{H}/\Gamma$. By Proposition 7.15, we can find a complementary $\mathbb{R}$-local system $\mathbb{U}'$ such that

$$R^1\phi_*(\mathbb{R}) = \mathbb{U} \oplus \mathbb{U}'.$$

This splitting defines a projector $\Psi \in \mathrm{End}(R^1\phi_*(\mathbb{R}))$ with image $\mathbb{U}$. We again use rigidity [PS03, Corollary 12] to show that Ψ is of bidegree $(0,0)$, i. e. a morphism of VHS. We apply Lemma 9.4 b) to the fiber over a point $c \in \mathcal{C}$ corresponding to X to deduce that Ψ_c is a morphism of Hodge structures. Hence by rigidity, Ψ is a morphism of VHS. Let $\mathcal{V} =$

$R^1\phi_*(\mathbb{R}) \otimes_{\mathbb{R}} \mathcal{O}_{\mathcal{C}}$ and let $\mathcal{V}^{1,0} = \phi_*\Omega_{\mathcal{X}/\mathcal{C}}$ be the non-trivial step of the filtration of the VHS. Its image

$$\Psi(\mathcal{V}^{1,0}) = \mathcal{V}^{1,0} \cap (\mathbb{U} \otimes_{\mathbb{R}} \mathcal{O}_{\mathcal{C}}) =: \mathcal{U}^{1,0}$$

is a holomorphic subbundle of $\mathcal{U} = \mathbb{U} \otimes_{\mathbb{R}} \mathcal{O}_{\mathcal{C}}$. To conclude that it defines a VHS on $\mathbb{U}$ we show that the filtration $\mathcal{U}^{1,0} \subset \mathcal{U}$ induces an $\mathbb{R}$-Hodge structure in each fiber. To this end, let $c \in \mathcal{C}$ be arbitrary and consider $\Psi_c : \mathcal{V}_c \to \mathcal{V}_c$, the map induced by Ψ in the fiber. It is a morphism of Hodge structures and induces a filtration $\mathrm{Im}(\Psi_c) \cap \mathcal{V}_c^{1,0} \subset \mathrm{Im}(\Psi_c)$ which is a (sub-)Hodge structure by [Voi02, Corollary 7.24]. Now $\mathrm{Im}(\Psi_c) \cap \mathcal{V}_c^{1,0} = \mathcal{U}_c^{1,0}$, so $\mathbb{U}$ carries a VHS. We polarize the VHS on $\mathbb{U}$ by pulling back the polarization on $R^1\phi_*(\mathbb{R})$ via the inclusion $\mathbb{U} \subset R^1\phi_*(\mathbb{R})$. ■

Theorem 9.7 *Let (X, ω) be a Veech surface, and let $G \leqslant \mathrm{Aut}(X, \omega)$. Consider an irreducible character χ, occuring in the decomposition* (9.1) *of $\Omega^1_X(X)$ for G. Let $V_\chi \subset \Omega^1_X(X)$ be its isotypic component. If*

- *(i) χ is complex-valued, and*
- *(ii) the $\mathbb{R}[G]$-module $W \subset H^1(X, \mathbb{R})$ associated with $V_\chi \oplus \overline{V_\chi}$ is an isotypic component of the decomposition* (9.3),

there is a finite index subgroup $\Gamma \leqslant \mathrm{Aff}(X, \omega)$, which acts through unitary matrices on $V_\chi \oplus \overline{V_\chi}$.

Proof: Let W be the $\mathbb{R}$-form of $V_\chi \oplus \overline{V_\chi}$, which is given by Lemma 9.4. Since W is an isotypic component, it follows from Proposition 9.5 that there exists a finite-index subgroup $\Gamma' \leqslant \mathrm{Aff}(X, \omega)$ fixing W. Thus it also fixes $W \otimes_{\mathbb{R}} \mathbb{C} = V_\chi \oplus \overline{V_\chi}$. Since V_χ and $\overline{V_\chi}$ belong to different isotypic components of the decomposition of $H^1(X, \mathbb{C})$, again by Proposition 9.5 there is a finite-index subgroup $\Gamma \leqslant \Gamma'$ fixing both V_χ and $\overline{V_\chi}$. In particular, for $\alpha, \beta \in V_\chi \subset \Omega^1_X(X)$, and every $f \in \Gamma$, we have

$$\begin{aligned} H((f^{-1})^*\alpha, (f^{-1})^*\beta) &= \int_X (f^{-1})^*\alpha \wedge *((f^{-1})^*\beta) = i \int_X f \cdot \alpha \wedge \overline{(f^{-1})^*\beta} \\ &= i \int_X (f^{-1})^*\alpha \wedge (f^{-1})^*\overline{\beta} = i \int_X \alpha \wedge \overline{\beta}\,, \end{aligned}$$

since f is symplectic. So f preserves the Hodge inner product, hence it acts on V_χ as a subgroup of the unitary group of H, and the same holds for $\overline{V_\chi}$. ∎

Theorem 9.7 and Proposition 8.17 together imply

Corollary 9.8 *In the situation of Theorem 9.7, the Lyapunov exponents associated with W are zero.*

Question 9.9 Will the assertion of Theorem 9.7 stay true, if we replace (i) by the condition that the irreducible character associated with W be reducible when tensoring with $\mathbb{C}$?

Chevalley-Weil Formula

Let $f : X \to Y$ be a Galois covering between compact Riemann surfaces, and let G be its Galois group. The Chevalley-Weil formula gives precise information about the number of times that a given irreducible character of G occurs in the decomposition of $\Omega^1_X(X)$ into irreducible G-modules.

Let $n = |G|$, and choose a primitive n-th root of unity ζ_n. For $d \mid n$, let $\zeta_d = \zeta_n^{n/d}$.

Let $B \subset Y$ be the set of branch points of f. For each $b \in B$, the stabilizer group $G_a \leqslant G$ of $a \in f^{-1}(b)$ is a non-trivial cyclic group; if $a' \in f^{-1}(b)$ is another ramification point, then G_a and $G_{a'}$ are conjugate in G. For every $b \in B$, fix the conjugacy class $C_b = [g_a]$ of a generator g_a of G_a. (C_b does not depend on the choice of $a \in f^{-1}(b)$.) Moreover let $e_b = |G_a|$ be the ramification index at $a \in f^{-1}(b)$.

Fix an irreducible representation (ρ, V) of G with character χ, and a point $b \in B$, and let $a \in f^{-1}(b)$. The restriction $\mathrm{Res}^G_{G_a}(V)$ to G_a decomposes as a sum of 1-dimensional characters of G_a. We use the isomorphism

$$\psi : \mathrm{Hom}(G_a, \mathbb{C}^\times) \to \mathbb{Z}/(e_b)\ , \quad \eta \mapsto \alpha$$

where $\eta(g_a) = \zeta_{e_b}^\alpha$. Let $N_{b,\alpha}$ be the number of times, $\psi^{-1}(\alpha)$ appears in the decomposition of $\mathrm{Res}^G_{G_a}(V)$. $N_{b,\alpha}$ is equal to the number of eigenvalues of $\rho(g_a)$ that are equal to $\zeta_{e_b}^{\alpha}$, counted with multiplicities. Again for $\alpha \in \mathbb{Z}/(e_b)$ the number $N_{b,\alpha}$ only depends on b and the chosen conjugacy class C_b.

One more notation: For $q \in \mathbb{R}$, let $\langle q \rangle = q - [q] \in [0,1)$ denote its fractional part.

Theorem 9.10 ([Wei35])
Let $f : X \to Y$ be a Galois covering between compact Riemann surfaces with Galois group G, and let B, e_b, $N_{b,\alpha}$ be defined as above. Let χ be an irreducible character of G of degree d_χ. Then the multiplicity of χ in the representation $\rho : G \to \mathrm{GL}(\Omega^1_X(X))$, $g \mapsto (g^{-1})^*$ is given by

$$\nu_\chi = d_\chi(g(Y)-1) + \sum_{b\in B} \sum_{\alpha\in\mathbb{Z}/(e_b)} \left(N_{b,\alpha}\left\langle -\frac{\alpha}{e_b}\right\rangle\right) + \sigma \;, \qquad (9.4)$$

where $\sigma = 1$, if χ is the trivial representation, and $\sigma = 0$ otherwise.

Proposition 9.11 *Let $\widetilde{\mathbf{St}}_3$ be the origami from Proposition 3.2. The Galois covering $\pi : \widetilde{\mathbf{St}}_3 \to E$ induces a decomposition*

$$\Omega^1_{\widetilde{\mathbf{St}}_3}(\widetilde{\mathbf{St}}_3) = V_1 \oplus V_5 \oplus V_6 \oplus V_7 \oplus V_8 \oplus V_9 \oplus V_{10} \oplus V_{11} \oplus V_{12} \oplus V_{13} \oplus V_{14} \oplus V_{15}^{\oplus 2}$$

into a direct sum of irreducible $\mathbb{C}[K_3]$-modules where K_3 is the Galois group of π and where we number the irreducible representations according to Table A.1.

Proof: We apply the Chevalley-Weil formula to π. To begin with, a generator of the monodromy about the only branch point $\infty \in E$ is $c = (\tau^2, \tau^{-2}, \tau^2) \in [\tau,\tau,\tau]$. We have $e_b = 3$. For $i = 1, \ldots, 15$, let ρ_i be the representation associated with the irreducible character χ_i of K_3. For $\alpha \in \mathbb{Z}/(3)$, let $N^i_{b,\alpha} = N^i_\alpha$ be the number of occurencies of ζ_3^α as eigenvalue of $\rho_i(c)$. We write it as a vector $N^i = (N^i_0, N^i_1, N^i_2)$.

- For $i = 1, \ldots, 4$, the degree of χ_i is 1, and the trace of $\rho_i(c) = 1$. Therefore $N^i = (1,0,0)$.
- For $i = 5, \ldots, 10$, the degree of χ_i is 2, and the trace of $\rho_i(c) = -1 = \zeta_3 + \zeta_3^2$. Therefore $N^i = (0,1,1)$.

- For $i = 11, 12, 13$, the degree of χ_i is 4, and the trace of $\rho_i(c) = 1 = 1 + 1 + \zeta_3 + \zeta_3^2$, so $N^i = (2, 1, 1)$.
- The degree of χ_{14} is 4, and trace of $\rho_{14}(c) = -2 - 3\zeta_3 = 1 + 3\zeta_3^2$. Hence $N^{14} = (1, 0, 3)$.
- The degree of χ_{15} is also 4, and the trace of $\rho_{15}(c) = 1 + 3\zeta_3$. Hence $N^{15} = (1, 3, 0)$.

Now we can plug these values into (9.4). Noting that the first summand cancels out, we have

$$\nu_{\chi_i} = \frac{2}{3} N_1^i + \frac{1}{3} N_2^i + \sigma.$$

This yields the decomposition of $\Omega^1_{\widetilde{\mathbf{St}}_3}(\widetilde{\mathbf{St}_3})$

$$\Omega^1_{\widetilde{\mathbf{St}}_3}(\widetilde{\mathbf{St}_3}) = V_1 \oplus V_5 \oplus V_6 \oplus V_7 \oplus V_8 \oplus V_9 \oplus V_{10} \oplus V_{11} \oplus V_{12} \oplus V_{13} \oplus V_{14} \oplus V_{15}^{\oplus 2}.$$

Recall that by Proposition 3.2, $g(\widetilde{\mathbf{St}_3}) = 37$, which fits together with the sum of the dimensions of the isotypic components of V_i, $i = 1, \dots, 15$. ■

O. Bauer [Bau09, Satz 3.5.1] computed the dimension of the fixed part of the family of Jacobians over the Teichmüller curve of $\widetilde{\mathbf{St}_3}$ (see Remark 8.18). Its dimension is 12. One can check, using e. g. MAGMA and the output of the origami program, that the fixed part coincides precisely with the sum of the isotypic components of the two complex representations χ_{14}, χ_{15}. Unfortunately, one cannot apply Theorem 9.7, since the isotypic component in $H^1(X, \mathbb{R})$ is the $\mathbb{R}$-form of the sum of the two isotypic components. It would be interesting to know if one can improve the proposition, so as to yield some information in this case.

9.3 Rank 2

The easiest case of a non-trivial symplectic direct summand in the local system of a family of curves is one of rank 2.

In this section, let (X,ω) denote a fixed Veech surface of area 1. Let

$$\rho : \mathrm{Aff}(X,\omega) \to \mathrm{Sp}(H^1(X,\mathbb{Z}))\ , \quad f \mapsto (f^{-1})^*$$

be the representation from Example 6.6. Let $\Gamma \leqslant \mathrm{Aff}(X,\omega)$ be a finite index-subgroup, chosen as in Condition $(*)$ of Remark 5.4, so that $\mathcal{C} = \mathbb{H}/\Gamma$ is a finite cover of the Teichmüller curve associated with (X,ω). Let $\overline{\mathcal{C}}$ be its completion, and let $S = \overline{\mathcal{C}} \setminus \mathcal{C}$ be the finite set of cusps. Moreover, let $\phi : \mathcal{X} \to \mathcal{C}$ be the family over the Teichmüller curve.

By Proposition 7.16, the local system $\mathbb{V} = R^1\phi_*(\mathbb{Z}_{\mathcal{X}})$ carries a pVHS

$$\mathcal{V}^{1,0} = \phi_*\Omega^1_{\mathcal{X}/\mathcal{C}} \subset \mathcal{V} = \mathbb{V} \otimes_{\mathbb{Z}} \mathcal{O}_{\mathcal{C}},$$

with polarization Q. Let us also fix this notation for the rest of the section.

Remark 9.12 Consider the subspace

$$U_1 = \mathrm{span}\{\mathrm{Re}\,\omega, \mathrm{Im}\,\omega\} \subset H^1(X,\mathbb{R}).$$

a) U_1 is an $\mathrm{Aff}(X,\omega)$-invariant symplectic subspace of dimension 2.

b) The action of $f \in \mathrm{Aff}(X,\omega)$ with $\mathrm{D}(f) = A$ is given by the matrix $(A^{-1})^T$ with respect to the basis $\{\mathrm{Re}\,\omega,\ \mathrm{Im}\,\omega\}$.

c) The local system $\mathbb{U}_1$ associated with the action of Γ carries a sub-pVHS of $(\mathbb{V}_{\mathbb{R}}, \mathcal{V}^{1,0}, Q)$.

d) The positive Lyapunov exponent associated with U_1 is $\lambda_1 = 1$.

PROOF: a) follows from the equation $\mathrm{Re}\,\omega \wedge \mathrm{Im}\,\omega = \frac{i}{2}\omega \wedge \overline{\omega}$ in $H^1(X,\mathbb{C})$. For part b), consider the action of $f \in \mathrm{Aff}(X,\omega)$ in a local chart of ω. If $\mathrm{D}(f) = \left(\begin{smallmatrix} a & b \\ c & d \end{smallmatrix}\right)$, then

$$(f^{-1})^*\mathrm{d}x = \mathrm{d}(dx - by) = d\mathrm{d}x - b\mathrm{d}y$$

and similarly $(f^{-1})^*\mathrm{d}y = -c\mathrm{d}x + a\mathrm{d}y$, which proves the claim.

c) follows from Proposition 7.28. In the case of origamis, it also follows from Theorem 9.3, applied to the origami map $\pi : \mathbf{O} \to E$.

d) We have to evaluate

$$\lim_{t\to\infty} \frac{1}{t} \log \|v\|_{\varphi_{g_t}}$$

for $v \in U_1$. For $v = \operatorname{Re}\omega$,

$$\|\operatorname{Re}\omega\|_{\varphi_{g_t}} = \|(\varphi_{g_t}^{-1})^* \operatorname{Re}\omega\|_{g_t\cdot X} = e^{-t}\|\operatorname{Re}(g_t\cdot\omega)\|_{g_t\cdot X}.$$

Since $\|\operatorname{Re}(g_t\cdot\omega)\|_{g_t\cdot X} = \frac{i}{2}\int_X g_t\cdot\omega\wedge\overline{g_t\cdot\omega} = 1$, it follows that the Lyapunov exponent associated with $\operatorname{Re}\omega$ is -1. Analogously, one shows that the Lyapunov exponent for $\operatorname{Im}\omega$ is equal to 1. ■

Definition 9.13 The representation $(\rho\otimes_{\mathbb{Z}}\mathbb{R})_{|U_1}$ is called the *trivial subrepresentation* of ρ. We denote by $\mathcal{L}_1$ the line bundle obtained as the $(1,0)$-part in the Deligne extension of the vector bundle $\mathbb{U}_1\otimes_{\mathbb{R}}\mathcal{O}_{\mathcal{C}}$ to $\overline{\mathcal{C}}$.

Proposition 9.14 ([BM10b, Theorem 8.5])
Assume that we are given a Γ-invariant subspace $U\subset H^1(X,\mathbb{R})$ of dimension 2, whose associated local system $\mathbb{U}\subset\mathbb{V}_{\mathbb{R}}$ carries a sub-pVHS of $(\mathbb{V}_{\mathbb{R}},\mathcal{V}^{1,0},Q)$. Let $\mathcal{L}_U$ be the $(1,0)$-part in the Deligne extension of $\mathbb{U}\otimes_{\mathbb{R}}\mathcal{O}_{\mathcal{C}}$ to $\overline{\mathcal{C}}$. Then the non-negative Lyapunov exponent λ_U associated with U is given by

$$\lambda_U = \frac{\deg(\mathcal{L}_U)}{\deg(\mathcal{L}_1)}.$$

Note that this is a version of M. Kontsevich's formula for the sum of the Lyapunov exponents [Kon97].

The Kodaira-Spencer Map revisited

One way to compute the degree of the line bundle $\mathcal{L}_U$ of Proposition 9.14 is to bring oneself into the situation, where the Kodaira-Spencer map is an isomorphism.

Remark 9.15 Let again $\mathbb{L}\subset\mathbb{V}_{\mathbb{R}}$ be a local system, which carries a sub-pVHS of $(\mathbb{V},\mathcal{V}^{1,0},Q)$ of rank 2, and let $\mathcal{L}$ be the $(1,0)$-part in the Deligne extension of $\mathbb{L}\otimes_{\mathbb{R}}\mathcal{O}_{\mathcal{C}}$ to $\overline{\mathcal{C}}$. Let

$$\overline{\nabla}:\mathcal{L}\to\mathcal{L}^{\otimes\,-1}\otimes\Omega^1_{\overline{\mathcal{C}}}(S)$$

be the Kodaira-Spencer map as introduced in 7.27.

In the situation above, if $\overline{\nabla}$ is an isomorphism, then

$$2\deg \mathcal{L} = \deg(\Omega^1_{\overline{\mathcal{C}}}(S)) = 2g(\overline{\mathcal{C}}) - 2 + |S|.$$

Proposition 9.16 *In the case when*

a) $\phi : \mathcal{X} \to \mathcal{C}$ *is a finite cover of a Teichmüller curve, and* $\mathbb{L}$ *is the sub-local system associated with* U_1*, or*

b) $\phi : \mathcal{X} \to \mathcal{C}$ *is a quotient of the universal family of elliptic curves over* $\mathbb{H}$ *by the action of a torsion-free, finite-index subgroup of* $\mathrm{SL}_2(\mathbb{Z})$*, and* $\mathbb{L} = R^1\phi_*(\mathbb{Z}_{\mathcal{X}})$*,*

the Kodaira-Spencer map $\overline{\nabla}$ *is an isomorphism.*

PROOF: a) is shown in [Möl06, Lemma 3.1]. b) follows from a), since $\mathcal{C}$ is a finite cover of the Teichmüller curve associated with $(E, \mathrm{d}z)$ (which is identical with $\mathcal{T}_{1,1} / \Gamma_{1,1}$). ∎

Proposition 9.17 *Assume that we are given a subrepresentation of* ρ *of rank 2*

$$\rho_{|U}^{|\Gamma} = \rho_U : \Gamma \to \mathrm{Sp}(U),$$

i. e. U *is a* Γ*-invariant rank-2 submodule of* $H^1(X, \mathbb{Z})$*. Assume further that the local system* $\mathbb{U}$ *associated with* U *carries a sub-pVHS of* $(\mathbb{V}, \mathcal{V}^{1,0}, Q)$*.*

If $\mathrm{Im}(\rho_U) \leqslant \mathrm{Sp}(U) \cong \mathrm{SL}_2(\mathbb{Z})$ *is of finite index, then the non-negative Lyapunov exponent of* ρ_U *satisfies*

$$\lambda_U = \frac{\deg(p)\deg(\mathcal{L}^{\rho_U})}{\deg(\mathcal{L}_1)},$$

where $p : \mathcal{C} \to \mathrm{Im}(\rho_U)\backslash\mathbb{H}$ *is the period map, and* $\mathcal{L}^{\rho_U}$ *is the* $(1,0)$*-part in the Deligne extension of the canonical VHS on* $\mathrm{Im}(\rho_U)\backslash\mathbb{H}$ *to* $\overline{\mathrm{Im}(\rho_U)\backslash\mathbb{H}}$*.*

Before beginning the proof, we introduce some terminology. Let Γ be a torsion-free Fuchsian group. Call a map $f : \Gamma \times \mathbb{H} \to \mathbb{C}^\times$ a *factor of automorphy* if it satisfies

$$f(\gamma\delta, \tau) = f(\gamma, \delta \cdot \tau) f(\delta, \tau)$$

for all $\gamma, \delta \in \Gamma$ and $\tau \in \mathbb{H}$ and if $f(\gamma, \cdot)$ is holomorphic for every $\gamma \in \Gamma$. Here, Γ acts by Möbius transformations on $\mathbb{H}$. Under pointwise multiplication, the factors of automorphy form a group $Z^1(\Gamma, H^0(\mathbb{H}, \mathcal{O}^\times))$. The subgroup $B^1(\Gamma, H^0(\mathbb{H}, \mathcal{O}^\times))$ of elements of the form

$$(\gamma, \tau) \mapsto h(\gamma \cdot \tau) h(\tau)^{-1}$$

for holomorphic $h : \mathbb{H} \to \mathbb{C}^\times$ is called the *group of boundaries*, and the quotient is denoted by $H^1(\Gamma, H^0(\mathbb{H}, \mathcal{O}^\times))$. It can be identified with the first cohomology group of Γ with values in $H^0(\mathbb{H}, \mathcal{O}^\times)$. There is a functorial isomorphism between the group $H^1(\mathbb{H}/\Gamma, \mathcal{O}^\times)$ of isomorphism classes of line bundles on $\mathbb{H}/\Gamma$ and $H^1(\Gamma, H^0(\mathbb{H}, \mathcal{O}^\times))$ (see e. g. [BL04, Appendix B]).

PROOF: To begin with, we give names to the projections $u : \mathbb{H} \to \mathbb{H}/\Gamma$ and $\tilde{u} : \mathbb{H} \to \rho_U(\Gamma)\backslash\mathbb{H}$. Let $\mathcal{U}^{1,0} \subset \mathbb{U} \otimes \mathcal{O}_{\mathcal{C}}$ be the sub-pVHS on $\mathbb{U}$, and let $\mathcal{E}$ be the universal line bundle on the period domain $\mathbb{H}$. It can be described as the sheaf of holomorphic sections for $V_E \to \mathbb{H}$, where

$$V_E = \left\{(\tau, v) \in \mathbb{H} \times \mathbb{C}^2 \mid v \in \mathbb{C} \cdot (\tau, 1)^T\right\}.$$

The period map $\tilde{p} : \mathbb{H} \to \mathbb{H}$ between the universal cover of $\mathcal{C}$ and the period domain was constructed to satisfy $\tilde{p}^* \mathcal{E} \cong u^* \mathcal{U}^{1,0}$. p is the holomorphic map induced by $\tilde{p}$, and we have a commutative diagram

$$\begin{array}{ccc} \mathbb{H} & \xrightarrow{\tilde{p}} & \mathbb{H} \\ {\scriptstyle u}\downarrow & & \downarrow{\scriptstyle \tilde{u}} \\ \mathbb{H}/\Gamma & \xrightarrow[p]{} & \rho_U(\Gamma)\backslash\mathbb{H} \end{array}$$

By Remark 7.10, an element $\gamma' = \left(\begin{smallmatrix} A & B \\ C & D \end{smallmatrix}\right) \in \rho(\Gamma)$ acts on $\mathbb{H}$ by its Möbius transformation action, and this action is induced by the action on V_E given by

$$(\tau', \lambda(\tau', 1)^T) \mapsto (\gamma' \cdot \tau', \lambda(C\tau' + D)(\gamma' \cdot \tau', 1)^T),$$

where $\lambda \in \mathbb{C}$. Let $\overline{\mathcal{E}}$ be the sheaf of sections of the quotient

$$\rho_U(\Gamma)\backslash V_E \to \rho_U(\Gamma)\backslash\mathbb{H}.$$

The line bundle $\mathcal{L}^{\rho_U}$ is the extension of $\overline{\mathcal{E}}$ to the completion. If we can show that

$$p^*\overline{\mathcal{E}} \cong \mathcal{U}^{1,0},$$

then the claim is proved. Since the isomorphism

$$H^1(\mathbb{H}/\Gamma, \mathcal{O}^\times) \to H^1(\Gamma, H^0(\mathbb{H}, \mathcal{O}^\times))$$

is functorial, we can instead prove that the group homomorphism

$$H^1(\rho(\Gamma), H^0(\mathbb{H}, \mathcal{O}^\times)) \to H^1(\Gamma, H^0(\mathbb{H}, \mathcal{O}^\times)), \quad f \mapsto f \circ (\rho \times p)$$

maps a factor of automorphy for $\overline{\mathcal{E}}$ to a factor of automorphy for $\mathcal{U}^{1,0}$. By the above discussion, a factor of automorphy corresponding to $\overline{\mathcal{E}}$ is given by

$$(\gamma', \tau') \mapsto f_{\overline{\mathcal{E}}}(\gamma', \tau') = (C\tau' + D).$$

A trivialization of $u^* \mathbb{U} \otimes \mathbb{C}$ is given by the sheaf of locally constant sections of

$$(\mathbb{H} \times U \to \mathbb{H}) \cong (\mathbb{H} \times \mathbb{C}^2 \to \mathbb{H}).$$

In this trivialization, the subbundle $u^* \mathcal{U}^{1,0}$ can be described as the sheaf of holomorphic sections of $V_U \to \mathbb{H}$ where

$$V_U = \{(\tau, v) \in \mathbb{H} \times \mathbb{C}^2 \mid v \in \mathbb{C} \cdot (p(\tau), 1)^T\}.$$

By Lemma 7.19 and its proof, $\gamma \in \Gamma$ acts on V_U by

$$(\tau, \lambda(p(\tau), 1)) \mapsto (\gamma \cdot \tau, \lambda \cdot (Cp(\tau) + D) \cdot (p(\gamma \cdot \tau), 1)),$$

where $\rho(\gamma) = \left(\begin{smallmatrix} A & B \\ C & D \end{smallmatrix}\right)$. Therefore, a factor of automorphy for $\mathcal{U}^{1,0}$ is given by

$$(\gamma, \tau) \mapsto Cp(\tau) + D = f_{\overline{\mathcal{E}}} \circ (\rho \times p)(\gamma, \tau).$$

This completes the proof. ∎

Going-up

In our computations, it is often impracticable to choose a subgroup Γ of $\mathrm{Aff}(X, \omega)$ fulfilling Condition $(*)$, as the index tends to be very large. To remedy this problem, we show that it suffices to carry out the computations for a finite index subgroup Δ of $\mathrm{Aff}(X,\omega)$, where a splitting of ρ is found.

Let Δ be a subgroup of $\mathrm{Aff}(X,\omega)$, and let $\sigma : \Delta \to \mathrm{Sp}(U) = \rho|^{\Delta}_{|U}$ be a symplectic subrepresentation of ρ of rank 2 such that $\sigma(\Delta)$ is a finite index subgroup of $\mathrm{Sp}(U) \cong \mathrm{SL}_2(\mathbb{Z})$. Assume again that the local system $\mathbb{U}$ associated with U carries a sub-pVHS of $(\mathbb{V}, \mathcal{V}^{1,0}, Q)$. W.l.o.g. we may assume $\Gamma \leqslant \Delta$. Let

$$\tilde{p} : \mathbb{H} \to \mathbb{H}$$

denote the period mapping associated with the pVHS on $\mathbb{U}$. By Lemma 7.19, $\tilde{p}$ induces holomorphic maps

$$p : \mathbb{H}/\Gamma \to \sigma(\Gamma)\backslash\mathbb{H} \quad \text{and} \quad q : \mathbb{H}/\Delta \to \sigma(\Delta)\backslash\mathbb{H}.$$

By our assumptions

$$\mathcal{D}_\Delta = \sigma(\Delta)\backslash\mathbb{H} \quad \text{and} \quad \mathcal{D}_\Gamma = \sigma(\Gamma)\backslash\mathbb{H}$$

are Riemann surfaces of finite type. Let $\overline{\mathcal{D}_\Delta}$ and $\overline{\mathcal{D}_\Gamma}$ be their completions and $S_\Delta = \overline{\mathcal{D}_\Delta} \setminus \mathcal{D}_\Delta$ respectively $S_\Gamma = \overline{\mathcal{D}_\Gamma} \setminus \mathcal{D}_\Gamma$ the set of cusps. Then we deduce from Proposition 9.17 that

Theorem 9.18 *In the situation above, the non-negative Lyapunov exponent associated with σ is*

$$\lambda_U = \frac{\deg(q)\,\mathrm{vol}(\sigma(\Delta)\backslash\mathbb{H})}{\mathrm{vol}(\mathbb{H}/\Delta)}$$

PROOF: By Proposition 9.17, and with the notations used there

$$\lambda_U = \frac{\deg(p)\deg(\mathcal{L}^\sigma)}{\deg(\mathcal{L}_1)}.$$

The Kodaira-Spencer map $\overline{\nabla}_\sigma$ for the VHS on $\sigma(\Gamma)\backslash\mathbb{H}$ is an isomorphism by Proposition 9.16 b). It follows that

$$2\deg(\mathcal{L}^\sigma_W) = 2g(\overline{\mathcal{D}_\Gamma}) - 2 + |S_\Gamma|.$$

By Proposition 9.16 a), we also have

$$2\deg(\mathcal{L}_1) = 2g(\overline{\mathcal{C}}) - 2 + |S|,$$

where S is the set of cusps of $\mathcal{C}$. By Gauß-Bonnet,

$$\operatorname{vol}(\sigma(\Gamma)\backslash \mathbb{H}) = 2\pi(2g(\overline{\mathcal{D}_\Gamma}) - 2 + |S_\Gamma|),$$

and

$$\operatorname{vol}(\mathbb{H}/\Gamma) = 2\pi(2g(\overline{\mathcal{C}}) - 2 + |S|).$$

Therefore,

$$\lambda_U = \frac{\deg(p)\operatorname{vol}(\sigma(\Gamma)\backslash \mathbb{H})}{\operatorname{vol}(\mathbb{H}/\Gamma)}.$$

We have a commutative diagram

$$\begin{array}{ccc} \mathbb{H}/\Gamma & \xrightarrow{p} & \sigma(\Gamma)\backslash \mathbb{H} \\ \downarrow & & \downarrow \\ \mathbb{H}/\Delta & \xrightarrow[q]{} & \sigma(\Delta)\backslash \mathbb{H} \end{array}$$

where the vertical arrows are induced by the two inclusions $\Gamma \leqslant \Delta$ and $\sigma(\Gamma) \leqslant \sigma(\Delta)$. This yields

$$\begin{aligned} \lambda_U &= \frac{\deg(p)(\sigma(\Delta):\sigma(\Gamma))\operatorname{vol}(\sigma(\Delta)\backslash \mathbb{H})}{(\Delta:\Gamma)\operatorname{vol}(\mathbb{H}/\Delta)} \\ &= \frac{\deg(q)\operatorname{vol}(\sigma(\Delta)\backslash \mathbb{H})}{\operatorname{vol}(\mathbb{H}/\Delta)}, \end{aligned}$$

which completes the proof. ■

Properties of Period Mappings in Rank 2

In order to be able to compute the Lyapunov exponents for the examples below, we recall some general results on Fuchsian groups and in particular finite-index subgroups of $\mathrm{SL}_2(\mathbb{Z})$.

First, let $\Gamma \leqslant \mathrm{PSL}_2(\mathbb{R})$ be a Fuchsian group of finite covolume. Then there is a bijection between orbifold points in $\Gamma \backslash \mathbb{H}$ and the conjugacy classes of non-trivial maximal finite cyclic subgroups of Γ and there is a bijection between the (finitely many) cusps of $\Gamma \backslash \mathbb{H}$ and the conjugacy classes of maximal parabolic subgroups of Γ (i. e. subgroups, where all elements are parabolic) (see [Kat92]).

If we specialise to $\Gamma \leqslant \mathrm{SL}_2(\mathbb{Z})$ of finite index, then we can assign to each parabolic element A a *width* $w(A) \in \mathbb{N}$, which is defined as the integer t, such that A is conjugate to $\pm\left(\begin{smallmatrix} 1 & \pm t \\ 0 & 1 \end{smallmatrix}\right)$. The *width* $w(c)$ of a cusp c is the width of a generator of the stabilizer of c, which is the same as the ramification index above $i\infty \in \mathrm{SL}_2(\mathbb{Z}) \backslash \mathbb{H}$.

The next proposition will later be used to compute the ramification indices and the degree of period mappings in rank 2 that are defined over the integers. To formulate it, consider the following setup. Let Γ, Δ be lattices in $\mathrm{SL}_2(\mathbb{R})$ and let $\rho : \Gamma \to \Delta$ be a group homomorphism. Let $\tilde{p} : \mathbb{H} \to \mathbb{H}$ be a ρ-equivariant holomorphic map, i. e.

$$\tilde{p}(\gamma \cdot z) = \rho(\gamma) \cdot \tilde{p}(z)$$

for all $\gamma \in \Gamma$ and all $z \in \mathbb{H}$. Let $p : \Gamma \backslash \mathbb{H} \to \Delta \backslash \mathbb{H}$ be the map induced by $\tilde{p}$, and let $\overline{p} : \overline{\Gamma \backslash \mathbb{H}} \to \overline{\Delta \backslash \mathbb{H}}$ be its extension to the completions of $\Gamma \backslash \mathbb{H}$, respectively $\Delta \backslash \mathbb{H}$.

Proposition 9.19 *In the above situation, assume in addition that* Γ, $\Delta \leqslant \mathrm{SL}_2(\mathbb{Z})$ *are two subgroups of finite index.*

a) Let γ *be a generator of the stabilizer of a cusp* s *of* Γ*, and assume that* $\rho(\gamma)$ *is parabolic. Let* t *be its fixed point and let* δ *be a generator of the stabilizer of* t *in* Δ*. Then* $\overline{p}(\Gamma \cdot s) = \Delta \cdot t$*, and the ramification index of* $\overline{p}$ *at* $\Gamma \cdot s$ *is given by*

$$e_{\Gamma \cdot s}(\overline{p}) = \frac{w(\rho(\gamma))}{w(\delta)}.$$

b) The degree of p *is given by* $\deg(p) = \sum_{c \in p^{-1}(c')} e_c(p)$ *for any cusp* c' *of* $\Delta \backslash \mathbb{H}$*.*

PROOF: a) The first assertion is shown in Lemma 9.20. From Lemma 9.22, it follows that $e_{\Gamma \cdot s}(\overline{p}) = (\langle\delta\rangle : \langle\rho(\gamma)\rangle)$. W. l. o. g. write $\delta = (z \mapsto z + u)$. Then $\rho(\gamma) = (z \mapsto z + v)$ with $u \mid v$, and $(\langle\delta\rangle : \langle\rho(\gamma)\rangle) = |v/u| = w(\rho(\gamma))/w(\delta)$. b) can be found in any textbook on compact Riemann surfaces, e. g. [For81, Theorem 4.24]. ∎

Lemma 9.20 *Let A be a parabolic element in Γ such that $B = \rho(A) \in \Delta$ is also parabolic. Let $s \in \mathbb{R} \cup \{\infty\}$ denote the fixed point of A, and let $t \in \mathbb{R} \cup \{\infty\}$ denote the fixed point of B. Then*

$$\Delta \cdot t = \overline{p}(\Gamma \cdot s).$$

PROOF: Without loss of generality, we may assume $s = t = \infty$. If not choose α, $\beta \in \mathrm{PSL}_2(\mathbb{R})$ such that $\alpha A \alpha^{-1}$ and $\beta B \beta^{-1}$ fix ∞. Then $\beta \circ \tilde{p} \circ \alpha^{-1} : \mathbb{H} \to \mathbb{H}$ is $\rho_1 : \alpha\Gamma\alpha^{-1} \to \beta\Delta\beta^{-1}$-equivariant. In particular, we can assume $A = (z \mapsto z + a)$ and $B = (z \mapsto z + b)$ for $a, b \in \mathbb{R} \setminus \{0\}$. For $R > 0$ let

$$U_R = \{z \in \mathbb{H} \mid \mathrm{Im}\,(z) > R\}.$$

$\Gamma \cdot U_R$ respectively $\Delta \cdot U_R$ are the elements of a neighborhood basis for $\Gamma \cdot s$ respectively $\Delta \cdot t$. Assume $\Delta \cdot t \neq \overline{p}(\Gamma \cdot s)$. Then we can choose disjoint open neighborhoods of both points since $\overline{\Delta \backslash \mathbb{H}}$ is Hausdorff. As $\overline{p}$ is open, the image of a neighborhood basis for $\Gamma \cdot s$ is a neighborhood basis for $\overline{p}(\Gamma \cdot s)$. Thus we can find R_1, $R_2 > 0$ such that

$$p(\Gamma \cdot U_{R_1}) \cap \Delta \cdot U_{R_2} = \emptyset.$$

On the other hand the Schwarz lemma implies that $\tilde{p}$ does not increase hyperbolic distances, so for every $z \in \mathbb{H}$

$$d_{\mathbb{H}}(z, A(z)) = d_{\mathbb{H}}(z, z + a) \geq d_{\mathbb{H}}(f(z), B(f(z))) = d_{\mathbb{H}}(f(z), f(z) + b).$$

Since $d_{\mathbb{H}}(z, z + a) \to 0$ if and only if $\mathrm{Im}\,(z) \to \infty$, it follows that for every $R' > 0$ there is $R'' > 0$ such that $\tilde{p}(U_{R'}) \subseteq U_{R''}$, and that $R'' \to \infty$, if $R' \to \infty$. Let $\tau \in U_{R_1}$. By increasing R_1 if necessary, we can achieve $\tilde{p}(\tau) \in U_{R_2}$. But then

$$p(\Gamma \cdot \tau) = \Delta \cdot \tilde{p}(\tau) \in p(\Gamma \cdot U_{R_1}) \cap \Delta \cdot U_{R_2}$$

contradicting disjointness. ∎

Lemma 9.21 *Let $\mathbb{D}^* = \mathbb{D}\backslash\{0\}$ be the punctured unit disk in $\mathbb{C}$, and let $f : \mathbb{D}^* \to \mathbb{D}^*$, $z \mapsto z^k$ (with $k \in \mathbb{N}$). Let $u : \mathbb{H} \to \mathbb{D}^*$, $z \mapsto \exp(2\pi i z)$ be a universal cover and let $\gamma : \mathbb{H} \to \mathbb{H}$, $z \mapsto z+1$. Then f lifts to $\tilde{f} : \mathbb{H} \to \mathbb{H}$ and $\tilde{f}(\gamma \cdot z) = \gamma^k \cdot \tilde{f}(z)$.*

Proof: From $u \circ \tilde{f} = f \circ u$, we deduce $\tilde{f} = kz + m$, for some $m \in \mathbb{Z}$. Then $\tilde{f}(\gamma \cdot z) = k(z+1) + m = \tilde{f}(z) + k = \gamma^k \cdot \tilde{f}(z)$. ■

Lemma 9.22 *Let γ be a generator of a maximal parabolic subgroup in Γ with fixed point s. Assume that $\rho(\gamma)$ is also parabolic with fixed point t, and let $\delta \in \Delta$ be a generator of the stabilizer of t. Then $\overline{p}(\Gamma \cdot s) = \Delta \cdot t$, and*

$$e_{\Gamma\cdot s}(\overline{p}) = (\langle\delta\rangle : \langle\rho(\gamma)\rangle).$$

Proof: Again, we can assume $s = t = \infty$, and $\gamma : z \mapsto z \pm 1$, $\delta : z \mapsto z \pm 1$. Let U_R be defined as in the proof of Lemma 9.20. For large R, two points in U_R are identified by Γ, if and only if they are identified by an element in $\langle\gamma\rangle$. So the map $\langle\gamma\rangle\backslash U_R \to \Gamma\backslash\mathbb{H}$ is injective, and the same holds for the map $\langle\delta\rangle\backslash U_R \to \Delta\backslash\mathbb{H}$. There exist charts about $\Gamma \cdot s$ in $\overline{\Gamma\backslash\mathbb{H}}$ and $\Delta \cdot t$ in $\overline{\Delta\backslash\mathbb{H}}$ such that $\overline{p}$, expressed in these charts, is of the form $z \mapsto z^k$, where $k = e_{\Gamma\cdot s}(\overline{p}) \in \mathbb{N}$ is the ramification index. Choose R large enough to assert $\tilde{p}(U_R) \subset U_R$. Then we have a commutative diagram

$$\begin{array}{ccc} \langle\gamma\rangle\backslash U_R & \longrightarrow & \mathbb{D}^* \\ p\big\downarrow & & \big\downarrow z \mapsto z^k \\ \langle\delta\rangle\backslash U_R & \longrightarrow & \mathbb{D}^* \end{array}$$

and U_R, which is biholomorphic to $\mathbb{H}$, is a universal cover of $\mathbb{D}^*$ and $\tilde{p}_{|U_R}$ is a lift of $z \mapsto z^k$. Because of the uniqueness of lifts and the universal cover, we are reduced to the situation of Lemma 9.21. γ and δ are mapped to generators of $\mathrm{Deck}(U_R/\mathbb{D}^*)$ via the isomorphism of U_R with $\mathbb{H}$, and $\rho(\gamma)$ becomes δ^k. Hence $(\langle\delta\rangle : \langle\rho(\gamma)\rangle) = k$. ■

9.4 Examples

In this section we carry out the computation of a splitting of the monodromy action and of the Lyapunov exponents for some examples. Our computations depend on the fact that we are able to fully decompose the monodromy action into pieces of rank 2.

The three examples that we discuss are origamis, which are derived from $\widetilde{\mathbf{St}}_3$ in the sense that they correspond to subgroups of the Galois group. They are connected by covering maps, coming from inclusions of the respective subgroups, as shown in Figure 9.9.

In this section, contrary to our convention and as a courtesy to at least one of my advisors, the composition $\alpha\beta$ of paths in the fundamental group of an origami is the path obtained by *first* running through α and *then* running through β.

To setup the notations, we first fix generators of $\mathrm{SL}_2(\mathbb{Z})$

$$S = \begin{pmatrix} 0 & -1 \\ 1 & 0 \end{pmatrix} \quad \text{and} \quad T = \begin{pmatrix} 1 & 1 \\ 0 & 1 \end{pmatrix}.$$

For $N \in \mathbb{N}$, let $\Gamma(N) \leqslant \mathrm{SL}_2(\mathbb{Z})$ be the kernel of the group homomorphism

$$\mathrm{SL}_2(\mathbb{Z}) \to \mathrm{SL}_2(\mathbb{Z}/(N))$$

induced by reduction mod N of the entries of the matrices. Recall that

$$\Gamma(2) = \left\langle A_1 = \begin{pmatrix} 1 & 2 \\ 0 & 1 \end{pmatrix}, A_2 = \begin{pmatrix} 1 & 0 \\ 2 & 1 \end{pmatrix}, -I_2 \right\rangle.$$

We also fix this generating set.

The Origami $\mathbf{L}_{2,2}$

$\mathbf{L}_{2,2}$, which is the same as $\mathbf{St}_3$, is the simplest origami, which is not of genus 1. Its $\mathrm{SL}_2(\mathbb{Z})$-orbit is given in Figure 9.1. It has yet been served as a toy model for many authors (see [Sch05a], [Kre10], [HL06]). The equation for its family of curves has been given by M. Möller [Möl05b, Proposition 4.1].

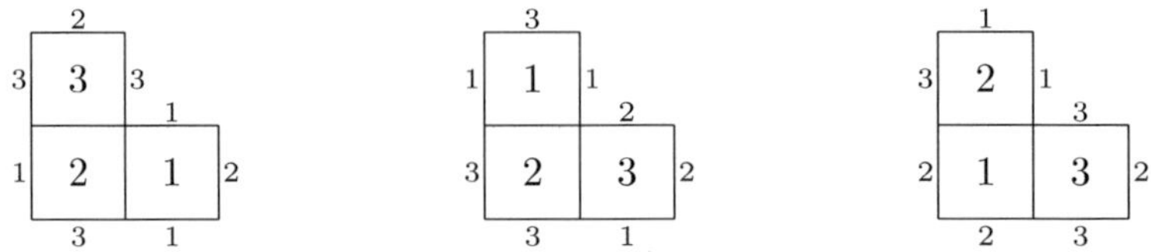

Figure 9.1: *The origamis $\mathbf{L}_{2,2}$, $T \cdot \mathbf{L}_{2,2}$, and $S^{-1}T^{-1} \cdot \mathbf{L}_{2,2}$ (from left to right).*

Its Veech group is

$$\Gamma_\Theta = \left\langle T^2 = \begin{pmatrix} 1 & 2 \\ 0 & 1 \end{pmatrix}, S = \begin{pmatrix} 0 & -1 \\ 1 & 0 \end{pmatrix}, TSTS^{-1}T^{-1} = \begin{pmatrix} 0 & 1 \\ -1 & 2 \end{pmatrix} \right\rangle,$$

which is an index 3-subgroup of $\mathrm{SL}_2(\mathbb{Z})$ with right cosets $\Gamma_\Theta, \Gamma_\Theta \cdot T$, and $\Gamma_\Theta \cdot (TS)$. In particular, the origami curve of $\mathbf{L}_{2,2}$ has genus 0 and 2 cusps, one of width 2, stabilized by T^2 and one of width 1, stabilized by $TSTS^{-1}T^{-1}$ (see [Kre10, Example 2.6]). Note that the two cusps of Γ_Θ have different images in the moduli space as can be seen from their stable graphs. As one quickly checks, there are no non-trivial translations, so we have $\mathrm{Aff}(\mathbf{L}_{2,2}) \cong \Gamma(\mathbf{L}_{2,2})$.

As $\mathbf{L}_{2,2}$ is in the stratum $\Omega\mathcal{M}_2(2)$, we know from Proposition 8.15 that its non-trivial Lyapunov exponent is $\frac{1}{3}$. Moreover, the representation $\rho : \mathrm{Aff}(\mathbf{L}_{2,2}) \to H^1(\mathbf{L}_{2,2}, \mathbb{Q})$ decomposes into two symplectic subrepresentations $(\rho_i, V_i)_{i=1,2}$. We let (ρ_1, V_1) denote the trivial subrepresentation.

Example 9.23 We illustrate how to compute the action of $\mathrm{Aff}(\mathbf{L}_{2,2})$ on the first homology (respectively first cohomology) of $\mathbf{L}_{2,2}$. Let $\pi : \mathbf{L}_{2,2} \to E$ be the origami cover, and let ${\mathbf{L}_{2,2}}^* = \mathbf{L}_{2,2} \setminus \pi^{-1}(\infty)$. Then ${\mathbf{L}_{2,2}}^*$ is homotopy equivalent to the 4-valent graph $\mathcal{G}(\mathbf{L}_{2,2})$ (see Remark 2.11). If we choose a maximal spanning tree of $\mathcal{G}(\mathbf{L}_{2,2})$, then by Seifert-Van Kampen the set of non-tree edges is in bijection with a free generating set of $\pi_1(\mathcal{G}(\mathbf{L}_{2,2}), *_1)$ (where $*_1$ is the center of the square labeled by 1). Fix the set of non-tree edges $\{t_i\}_{i=1}^4$ indicated in Figure 9.2. To apply affine homeomorphisms to $\mathbf{L}_{2,2}$, it will be convenient to look at the image

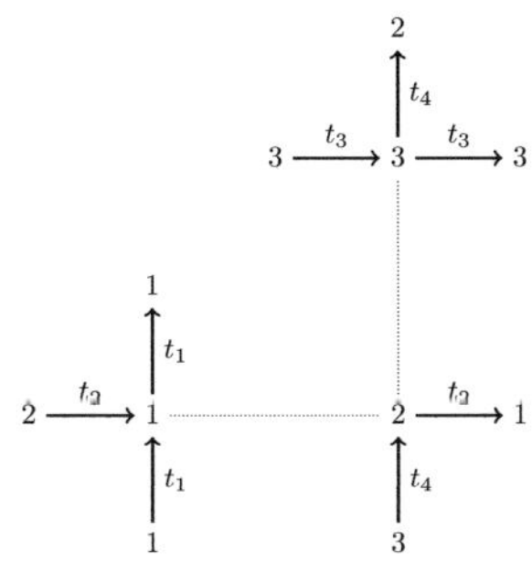

Figure 9.2: *The graph of the origami* $\mathbf{L}_{2,2}$ *with* π_1*-basis*

$H(\mathbf{L}_{2,2})$ of $\pi_1(\mathbf{L}_{2,2}{}^*, *_1) \cong \pi_1(\mathcal{G}(\mathbf{L}_{2,2}), *_1)$ inside $\pi_1(E^*, e) = \langle x, y\rangle$. Here $E^* = E \setminus \{\infty\}$, $e \in E$ is the center of the square, and x (respectively y) is the horizontal (vertical) loop on E^*. We have

$$t_1 \mapsto y \qquad t_2 \mapsto x^2$$
$$t_3 \mapsto xyxy^{-1}x^{-1} \qquad t_4 \mapsto xy^2x^{-1}$$

Then $\pi_1(\mathbf{L}_{2,2}, *_1) \cong H(\mathbf{L}_{2,2}) = \langle y, x^2, xyxy^{-1}x^{-1}, xy^2x^{-1}\rangle$. We choose lifts of the generators of Γ_Θ to $\mathrm{Aut}^+(F(x,y))$ that stabilize $H(\mathbf{L}_{2,2})$:

$$\begin{aligned}\varphi_{T^2}(x,y) &= (x, x^2y)\\ \varphi_S(x,y) &= (x^{-1}yx, (x^{-1}y)x^{-1}(x^{-1}y)^{-1})\\ \varphi_{TSTS^{-1}T^{-1}}(x,y) &= (xyxy^{-1}x^{-1}y^{-1}x^{-1}, xyxyx^{-1}).\end{aligned}$$

We carry out the following steps only for T^2. Applying φ_{T^2} to $H(\mathbf{L}_{2,2})$ transforms the generators of $H(\mathbf{L}_{2,2})$ into

$$x^2y\ ,\ x^2\ ,\ x^3yxy^{-1}x^{-3}\ , \quad \text{and} \quad x^3yx^2yx^{-1}.$$

By going back from $H(\mathbf{L}_{2,2})$ to $\pi_1(\mathcal{G}(\mathbf{L}_{2,2}), *_1)$, we find that T^2 acts on the generators $\{t_i\}_{i=1}^4$ by

$$t_1 \mapsto t_2t_1 \qquad t_2 \mapsto t_2$$
$$t_3 \mapsto t_2t_3t_2^{-1} \qquad t_4 \mapsto t_2t_3^2t_4$$

After projecting to $H_1(\mathbf{L}_{2,2}{}^*, \mathbb{Z})$, we obtain a basis $\{\overline{t_i}\}_{i=1}^4$, and the action of T^2 on $H_1(\mathbf{L}_{2,2}{}^*, \mathbb{Z})$ with respect to this basis is given by

$$\tilde{A}_{T^2} = \begin{pmatrix} 1 & 0 & 0 & 0 \\ 1 & 1 & 0 & 1 \\ 0 & 0 & 1 & 2 \\ 0 & 0 & 0 & 1 \end{pmatrix}.$$

In the same way, we obtain the actions of the remaining generators

$$\tilde{A}_S = \begin{pmatrix} 0 & 0 & 1 & 0 \\ 0 & 0 & 0 & -1 \\ -1 & 0 & 0 & 0 \\ 0 & 1 & 0 & 0 \end{pmatrix}, \quad \tilde{A}_{TSTS^{-1}T^{-1}} = \begin{pmatrix} 0 & 0 & -1 & 2 \\ 0 & 0 & 0 & 1 \\ 1 & 0 & 0 & 0 \\ 1 & -1 & 0 & 1 \end{pmatrix}.$$

We modify our basis in order to obtain a symplectic basis by using surface normalization as described in [Sti80, Section 1.3]. Let

$$s_1 = t_1 \ , \ s_2 = t_1^{-1} t_2^{-1} t_1 \ , \ s_3 = t_1^{-1} t_3 t_1 \ , \ s_4 = t_1^{-1} t_4.$$

Then $\{s_i\}_{i=1}^4$ is another basis of $\pi_1(\mathcal{G}(\mathbf{L}_{2,2}), *_1)$, and its projection to $H_1(\mathbf{L}_{2,2}{}^*, \mathbb{Z})$ is a symplectic basis for the intersection form i of Example 4.2. The matrix of i with respect to $\{s_i\}_{i=1}^4$ is given by

$$\Omega = \begin{pmatrix} 0 & 1 & 0 & 0 \\ -1 & 0 & 0 & 0 \\ 0 & 0 & 0 & 1 \\ 0 & 0 & -1 & 0 \end{pmatrix}.$$

Note that $\{s_i\}_{i=1}^4$ is already a basis for $H_1(\mathbf{L}_{2,2}, \mathbb{Z})$, while in strata with more zeros, we obtain a symplectic basis of the absolute homology of the closed surface, extended by loops about the zeros. After base change, we end up with symplectic matrices

$$A_{T^2} = \begin{pmatrix} 1 & 0 & 0 & 0 \\ -1 & 1 & 0 & 0 \\ 0 & 0 & 1 & 2 \\ 0 & 0 & 0 & 1 \end{pmatrix}, \ A_S = \begin{pmatrix} 0 & -1 & 1 & 0 \\ 0 & 0 & 0 & 1 \\ -1 & 0 & 0 & 1 \\ 0 & -1 & 0 & 0 \end{pmatrix},$$

$$A_{TSTS^{-1}T^{-1}} = \begin{pmatrix} 1 & 1 & -1 & 2 \\ 0 & 0 & 0 & -1 \\ 1 & 0 & 0 & -1 \\ 1 & 1 & 0 & 0 \end{pmatrix},$$

which represent the action of $\mathrm{Aff}(\mathbf{L}_{2,2})$ on $H_1(\mathbf{L}_{2,2}, \mathbb{Z})$. The dual action of $\mathrm{Aff}(\mathbf{L}_{2,2})$ on $H^1(\mathbf{L}_{2,2}, \mathbb{Z})$ with respect to the dual basis $\{s_i^*\}_{i=1}^4$ is therefore given by conjugating[1] the above matrices by Ω:

$$\Omega A_{T^2}\Omega^{-1} = \begin{pmatrix} 1 & 1 & 0 & 0 \\ 0 & 1 & 0 & 0 \\ 0 & 0 & 1 & 0 \\ 0 & 0 & -2 & 1 \end{pmatrix}, \quad \Omega A_S \Omega^{-1} = \begin{pmatrix} 0 & 0 & 1 & 0 \\ 1 & 0 & 0 & 1 \\ -1 & 0 & 0 & 0 \\ 0 & -1 & -1 & 0 \end{pmatrix},$$

$$\Omega A_{TSTS^{-1}T^{-1}}\Omega^{-1} = \begin{pmatrix} 0 & 0 & -1 & 0 \\ -1 & 1 & -2 & -1 \\ 1 & -1 & 0 & 0 \\ 0 & 1 & 1 & 0 \end{pmatrix}.$$

Note that we can read off the subspace corresponding to the trivial representation. Consider the space in homology spanned by h and v, where h is the sum of all horizontal cycles, and v is the sum of all vertical cycles. Then the corresponding space in cohomology is the image of $\langle h, v\rangle$ under the isomorphism

$$H_1(\mathbf{L}_{2,2}, \mathbb{Z}) \to H^1(\mathbf{L}_{2,2}, \mathbb{Z}), \quad a \mapsto i(\cdot, a).$$

In our example, we have $h = \overline{t_3} + \overline{t_2}$ and $v = \overline{t_1} + \overline{t_4}$, which after base change to $\{s_i\}_{i=1}^4$ have the coordinate vectors $(0, -1, 1, 0)^T$ and $(2, 0, 0, 1)^T$. Therefore,

$$V_1 = \mathrm{span}\{-s_1^* - s_4^*, -2s_2^* + s_3^*\} \subset H^1(\mathbf{L}_{2,2}, \mathbb{Z})$$

is the trivial subrepresentation of $\rho : \mathrm{Aff}(\mathbf{L}_{2,2}) \to \mathrm{Sp}(H^1(\mathbf{L}_{2,2}, \mathbb{Z}))$, i. e.

$$V_1 \otimes_{\mathbb{Z}} \mathbb{R} = \mathrm{span}\{\mathrm{Re}\,\omega, \mathrm{Im}\,\omega\},$$

where ω is differential defining the translation structure.

[1] Note that $\Omega A \Omega^{-1} = (A^{-1})^T$ for symplectic matrices.

Remark 9.24 With respect to an appropriate basis,

$$w_1 = s_1^* - 2s_4^* \ , \quad w_2 = s_2^* + s_3^*$$

of $V_2 = V_1^\perp$, the action of the generators of Γ_Θ under the non-trivial representation $\rho_2 : \Gamma_\Theta \to \mathrm{Sp}(V_2) \cong \mathrm{SL}_2(\mathbb{Z})$ is given by

$$\rho_2(T^2) = \begin{pmatrix} 1 & 1 \\ 0 & 1 \end{pmatrix} \ , \quad \rho_2(S) = \begin{pmatrix} 0 & 1 \\ -1 & 0 \end{pmatrix} \ , \quad \rho_2(TSTS^{-1}T^{-1}) = \begin{pmatrix} 0 & -1 \\ 1 & -1 \end{pmatrix}$$

The Origami M

Next we consider the origami $\mathbf{M}$ given as in Figure 9.3, or equivalently by the two permutations

$$\sigma_x = (1,4,7)(2,3,5,6,8,9) \quad \text{and} \quad \sigma_y = (1,6,8,7,3,2)(4,9,5).$$

The $\mathrm{SL}_2(\mathbb{Z})$-orbit of $\mathbf{M}$ contains two more origamis $T \cdot \mathbf{M}$ and $S^{-1}T^{-1} \cdot \mathbf{M}$. They are depicted in Figures 9.4 and 9.5. Each of these 3 origamis is a 3-fold cover of precisely two origamis in $\mathrm{SL}_2(\mathbb{Z}) \cdot \mathbf{L}_{2,2}$ (see also Figure 9.9). For $\mathbf{M}$, these covering maps can be constructed from the action of σ_x and σ_y on the partitions

$$(\{1,4,7\},\{2,5,8\},\{3,6,9\}) \ , \ (\{1,3,8\},\{2,6,7\},\{4,5,9\}).$$

The first partition corresponds to $T \cdot \mathbf{L}_{2,2}$ with the squares 1, 4, 7 being sent to 1, the squares 2, 5, 8 being sent to 2 and the squares 3, 6, 9 being sent to 3, the second corresponds to $S^{-1}T^{-1} \cdot \mathbf{L}_{2,2}$ in an analogous way.

We collect some facts about $\mathbf{M}$. From Figure 9.3, we see that it belongs to the stratum $\Omega\mathcal{M}_4(2^3)$. Moreover, it is in the “odd” connected component.

Its only non-trivial automorphism is a hyperelliptic involution. It takes the square i, rotates it by π, and maps it to the square $\sigma(i)$, where

$$\sigma = (1,7)(2,3)(4)(5,9)(6,8).$$

One can compute its Veech group using [Sch04], and one finds that $\Gamma(\mathbf{M}) = \Gamma_\Theta$. As before there are two cusps; for the convenience of the reader, we

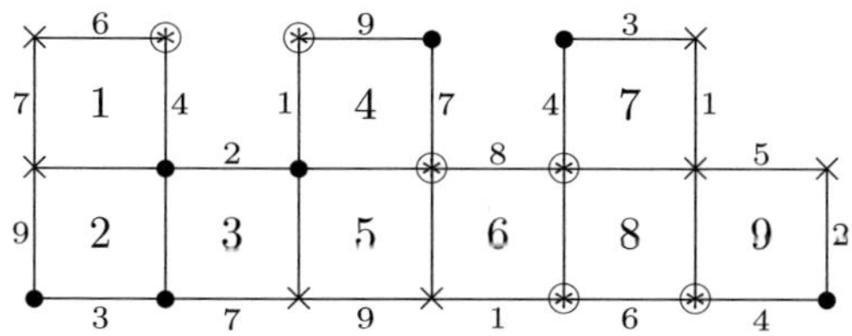

Figure 9.3: *The origami* **M**

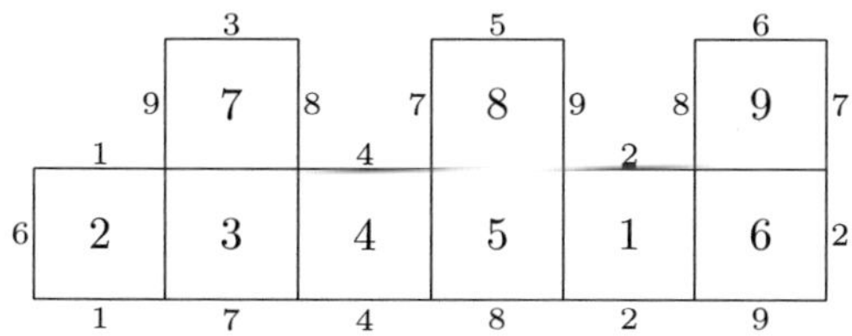

Figure 9.4: *The origami* $T \cdot \mathbf{M}$

describe the combinatorics of the stable curves in $\mathrm{Im}(j(\mathbf{M})) \cap \partial \mathcal{M}_4$. One, obtained e. g. by contracting the waist curves of horizontal cylinders of **M**, is a stable curve of genus 2 consisting of one irreducible component with 2 nodes. The other one, obtained e. g. by contracting the waist curves of horizontal cylinders in $S^{-1}T^{-1} \cdot \mathbf{M}$, consists of two irreducible components of genus 0 intersecting in two points. One of the components has two self-intersections, the other one has one.

In the following, we describe a splitting of the monodromy action of **M**. Let Γ be a finite-index subgroup of $\mathrm{Aff}(X, \omega)$ fulfilling Condition $(*)$, and let $\phi : \mathcal{X} \to \mathbb{H}/\Gamma$ be the family over the origami curve. Let $(\mathbb{V}, \mathcal{V}^{1,0}, Q)$ be the pVHS on $\mathbb{H}/\Gamma$ induced by ϕ.

Theorem 9.25 *The monodromy representation $\rho : \mathrm{Aff}(\mathbf{M}) \to H^1(\mathbf{M}, \mathbb{Z})$ restricted to $\Gamma(2)$ splits over $\mathbb{Q}$ into four symplectic subrepresentation $(\rho_i \otimes \mathbb{Q}, V_i \otimes \mathbb{Q})_{i=1}^4$. On the finite cover $\mathbb{H}/\Gamma$ of the origami curve $\mathcal{C}(\mathbf{M})$, each of the associated local systems carries a sub-pVHS whose period map-*

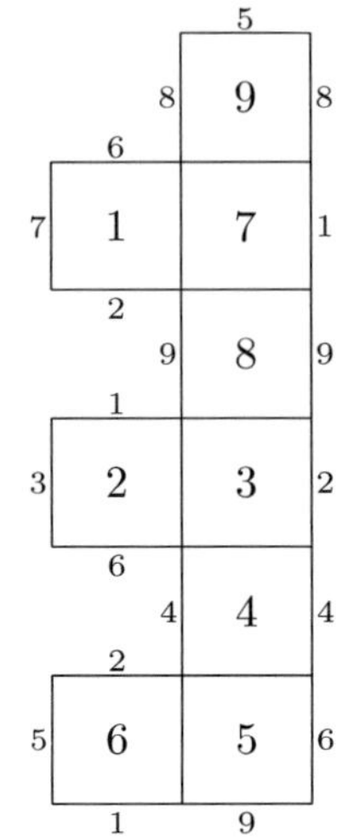

Figure 9.5: *The origami $S^{-1}T^{-1} \cdot \mathbf{M}$*

pings are denoted by p_i. Besides the trivial representation (ρ_1, V_1) there are

- *two representations (ρ_i, V_i), with $\rho_i = \rho_{|V_i}^{|\Gamma(2)}$, $i = 2, 3$, which are pullbacks of the non-trivial representations of $T \cdot \mathbf{L}_{2,2}$ and $S^{-1}T^{-1} \cdot \mathbf{L}_{2,2}$. They satisfy*

$$\mathrm{Im}\,(\rho_i) = \mathrm{Sp}(V_i) \cong \mathrm{SL}_2(\mathbb{Z}) \quad \textit{and} \quad \deg(p_i) = 2,$$

- *one representation (ρ_4, V_4) with $\rho_4 = \rho_{|V_4}$ that splits off over Γ_Θ. It satisfies*

$$\mathrm{Sp}(V_4) \cong \mathrm{SL}_2(\mathbb{Z}) \quad \textit{and} \quad \deg(p_4) = 1.$$

Proof: The computation proceeds as in Example 9.23. We indicate which choices we have made. First, a basis $\{t_i\}_{i=1}^{10}$ of $\pi_1(\mathcal{G}(\mathbf{M}))$ is shown in Figure 9.6. We find again a symplectic basis for $H_1(\mathbf{M}, \mathbb{Z})$ by surface normalization. Let $\{\overline{t_i}\}_{i=1}^{10}$ be the projection of $\{t_i\}_{i=1}^{10}$ to $H_1(\mathbf{M}, \mathbb{Z})$, and

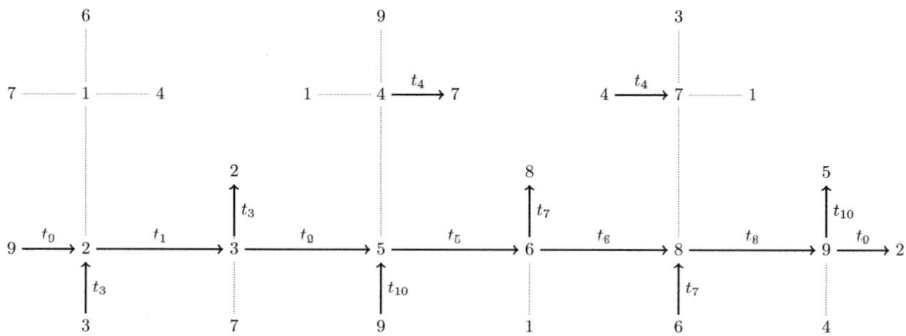

Figure 9.6: *The graph of the origami* **M** *with* π_1*-basis*

let

$$B = (b_{ij})_{i,j=1,\ldots,10} = \begin{pmatrix} 0 & 0 & 0 & 0 & 0 & 0 & 0 & 1 & 0 & 0 \\ 1 & -1 & 0 & -1 & 1 & 0 & 1 & 1 & 1 & 1 \\ 0 & 0 & 0 & 0 & 0 & 0 & -1 & 0 & 0 & 0 \\ 0 & 0 & 0 & 0 & 0 & 0 & 0 & 0 & 1 & 0 \\ 0 & 0 & 0 & 0 & 0 & 0 & 0 & 0 & 0 & 1 \\ 1 & 0 & -1 & -1 & 1 & 0 & 0 & 0 & 0 & 0 \\ -1 & 0 & 1 & 0 & 0 & 0 & 0 & 0 & 0 & 0 \\ 1 & 0 & -1 & -1 & 0 & 1 & 0 & 0 & 0 & -1 \\ 0 & 0 & -1 & 0 & -1 & 1 & 0 & 0 & -1 & -1 \\ 0 & 0 & 1 & 0 & 0 & -1 & 0 & 0 & 0 & 0 \end{pmatrix}.$$

Its inverse is

$$B^{-1} = \begin{pmatrix} 0 & 0 & 0 & -1 & 0 & -1 & -1 & 1 & -1 & 0 \\ 1 & -1 & -1 & 0 & 1 & 0 & 0 & 1 & -1 & 0 \\ 0 & 0 & 0 & -1 & 0 & -1 & 0 & 1 & -1 & 0 \\ 0 & 0 & 0 & -1 & -1 & -1 & -1 & 0 & -1 & -1 \\ 0 & 0 & 0 & -1 & -1 & 0 & 0 & 0 & -1 & -1 \\ 0 & 0 & 0 & -1 & 0 & -1 & 0 & 1 & -1 & -1 \\ 0 & 0 & -1 & 0 & 0 & 0 & 0 & 0 & 0 & 0 \\ 1 & 0 & 0 & 0 & 0 & 0 & 0 & 0 & 0 & 0 \\ 0 & 0 & 0 & 1 & 0 & 0 & 0 & 0 & 0 & 0 \\ 0 & 0 & 0 & 0 & 1 & 0 & 0 & 0 & 0 & 0 \end{pmatrix}.$$

We define a new basis $\{s_i\}_{i=1}^{10}$ by $s_j = \sum_{i=1}^{10} b_{ij}\overline{t_i}$. Note that the last two columns of B express loops around the punctures and will be neglected when closing the surface, i. e. when working with $\mathbf{M}$: s_9 is a positive loop about $\bullet$, and s_{10} is a negative loop about $\times$. The remaining set $\{s_i\}_{i=1}^{8}$ is a symplectic basis for the intersection form i. More precisely, the matrix Ω of i with respect to $\{s_i\}_{i=1}^{8}$ has four 2×2-blocks of $\left(\begin{smallmatrix} 0 & 1 \\ -1 & 0 \end{smallmatrix}\right)$ on the diagonal and zeros elsewhere. The action of $\mathrm{Aff}(\mathbf{M})$ on $H_1(\mathbf{M}, \mathbb{Z})$ in terms of generators with respect to the basis $\{s_i\}_{i=1}^{8}$ is given by

$$A_{T^2} = \begin{pmatrix} 1 & 1 & -2 & 1 & -1 & 2 & 0 & 0 \\ 1 & 0 & -1 & 0 & -1 & 1 & -2 & 0 \\ 1 & 1 & -1 & 1 & -1 & 1 & 0 & 0 \\ 0 & 2 & -3 & 2 & -1 & 3 & 2 & -1 \\ 1 & 1 & -3 & 1 & -1 & 3 & 1 & -1 \\ 1 & 1 & -1 & 1 & -1 & 1 & 1 & 0 \\ 1 & 0 & -1 & 0 & 0 & 0 & 0 & 0 \\ -1 & 0 & 1 & -1 & 1 & 0 & 0 & 0 \end{pmatrix},$$

$$A_S = \begin{pmatrix} -1 & 0 & 2 & 0 & 1 & -2 & 0 & 0 \\ -2 & 1 & 2 & 1 & 1 & -2 & 1 & 1 \\ -1 & 0 & 2 & 0 & 1 & -2 & 0 & 1 \\ 2 & -2 & 0 & -2 & 1 & 0 & -1 & -1 \\ 2 & -2 & 0 & -2 & 1 & 0 & 0 & 0 \\ 1 & -1 & 1 & -1 & 1 & -1 & 0 & 1 \\ -1 & 0 & 1 & 1 & -1 & 0 & 0 & 0 \\ 1 & 0 & -1 & 0 & 0 & 0 & 0 & 0 \end{pmatrix},$$

$$A_{(TS)T(TS)^{-1}} = \begin{pmatrix} 1 & 1 & -3 & 0 & 0 & 2 & 0 & 0 \\ -1 & 1 & -1 & -1 & 2 & 1 & 0 & 0 \\ -1 & 1 & -1 & 0 & 1 & 1 & 0 & 0 \\ 1 & 2 & -4 & 3 & -4 & 3 & 0 & 0 \\ -1 & 2 & -1 & 3 & -3 & 1 & 0 & 0 \\ -2 & 2 & -1 & 2 & -1 & 1 & 0 & 0 \\ 0 & 0 & 0 & 0 & 0 & 0 & 2 & 1 \\ 0 & 0 & 0 & 0 & 0 & 0 & -1 & 0 \end{pmatrix}.$$

The corresponding matrices for the action $\rho : \mathrm{Aff}(\mathbf{M}) \to \mathrm{Sp}(H^1(\mathbf{M}, \mathbb{Z}))$ are again obtained by conjugation with the intersection matrix Ω:

$$\Omega A_{T^2}\Omega^{-1} = \begin{pmatrix} 0 & -1 & 0 & 1 & 1 & 1 & 0 & 2 \\ -1 & 1 & -1 & -2 & -2 & -1 & 0 & 0 \\ 2 & 0 & 2 & 3 & 3 & 1 & -1 & -2 \\ -1 & 1 & -1 & -1 & -1 & -1 & 0 & 0 \\ 1 & -1 & 1 & 1 & 1 & 1 & 0 & -1 \\ -1 & 1 & -1 & -3 & -3 & -1 & 1 & 1 \\ 0 & 1 & -1 & -1 & 0 & -1 & 0 & 0 \\ 0 & 1 & 0 & -1 & 0 & 0 & 0 & 0 \end{pmatrix},$$

$$\Omega A_S\Omega^{-1} = \begin{pmatrix} 1 & 2 & 1 & -2 & -2 & -1 & 1 & -1 \\ 0 & -1 & 0 & 2 & 2 & 1 & 0 & 0 \\ -2 & -2 & -2 & 0 & 0 & -1 & -1 & 1 \\ 0 & -1 & 0 & 2 & 2 & 1 & -1 & 0 \\ -1 & -1 & -1 & -1 & -1 & -1 & 1 & 0 \\ 2 & 2 & 2 & 0 & 0 & 1 & 0 & 0 \\ 0 & -1 & 0 & 1 & 0 & 0 & 0 & 0 \\ 0 & -1 & -1 & 1 & 0 & -1 & 0 & 0 \end{pmatrix},$$

$$\Omega A_{TSTS^{-1}T^{-1}}\Omega^{-1} = \begin{pmatrix} 1 & 1 & -1 & 1 & 1 & -2 & 0 & 0 \\ -1 & 1 & 0 & -3 & -2 & 0 & 0 & 0 \\ 2 & -1 & 3 & 4 & 3 & 4 & 0 & 0 \\ -1 & -1 & 0 & -1 & -1 & 1 & 0 & 0 \\ 2 & 2 & 2 & 1 & 1 & 1 & 0 & 0 \\ -2 & -1 & -3 & -1 & -1 & -3 & 0 & 0 \\ 0 & 0 & 0 & 0 & 0 & 0 & 0 & 1 \\ 0 & 0 & 0 & 0 & 0 & 0 & -1 & 2 \end{pmatrix}.$$

To find the trivial representation, consider the cycles

$$h = \overline{t_1} + \overline{t_2} + \overline{t_4} + \overline{t_5} + \overline{t_6} + \overline{t_8} + \overline{t_9} \quad \text{and} \quad v = \overline{t_3} + \overline{t_7} + \overline{t_{10}}.$$

As in the previous example, we represent them in the basis $\{s_i\}_{i=1}^8$,

$$h = -2s_1 + s_2 - 2s_3 - 4s_4 - 3s_5 - 2s_6 + s_8$$

and

$$v = -s_1 - s_2 - 2s_4 - s_5 - s_6 - s_7.$$

and map them to $H^1(\mathbf{M}, \mathbb{Z})$ (with the isomorphism induced by i) to obtain

$$v_1^1 = s_1^* + 2s_2^* - 4s_3^* + 2s_4^* - 2s_5^* + 3s_6^* + s_7^*$$

and

$$v_2^1 = -s_1^* + s_2^* - 2s_3^* - s_5^* + s_6^* + s_8^*.$$

It follows that v_1^1 and v_2^1 span the trivial representation V_1.

Next, consider the inclusions

$$\varphi^* : H^1(T \cdot \mathbf{L}_{2,2}, \mathbb{Z}) \to H^1(\mathbf{M}, \mathbb{Z})$$

and

$$\varpi^* : H^1(S^{-1}T^{-1} \cdot \mathbf{L}_{2,2}, \mathbb{Z}) \to H^1(\mathbf{M}, \mathbb{Z})$$

given by the coverings $\varphi : \mathbf{M} \to T \cdot \mathbf{L}_{2,2}$, and $\varpi : \mathbf{M} \to S^{-1}T^{-1} \cdot \mathbf{L}_{2,2}$. We choose again bases of $H_1(T \cdot L22, \mathbb{Z})$ and $H_1(S^{-1}T^{-1} \cdot \mathbf{L}_{2,2}, \mathbb{Z})$; they are indicated in Figure 9.7. The matrices of φ_* and ϖ_* with respect to the t-bases (upstairs and downstairs) are

$$C_{\varphi_*} = \begin{pmatrix} 1 & -2 & -1 & 3 & 1 & 1 & 1 & -2 & 1 & 0 \\ 1 & 0 & 0 & 0 & 1 & 0 & 0 & 1 & 0 & 0 \\ 0 & 1 & 0 & 0 & 0 & 1 & 0 & 0 & 1 & 0 \\ 0 & 0 & 1 & 0 & 0 & 0 & 1 & 0 & 0 & 1 \end{pmatrix},$$

and

$$C_{\varpi_*} = \begin{pmatrix} 0 & 0 & 1 & 0 & -1 & 2 & 2 & -1 & 0 & 0 \\ -2 & 1 & 2 & 0 & 0 & 0 & 1 & 0 & 1 & 0 \\ 0 & 0 & 0 & 1 & 1 & 0 & 0 & 0 & 1 & 0 \\ 0 & 1 & 0 & 0 & -1 & 0 & 0 & -1 & 1 & 3 \end{pmatrix}.$$

Now, we change the basis upstairs to $\{s_i\}_{i=1}^{10}$, and obtain matrices w. r. t. the dual bases for the maps on cohomology (of the punctured surfaces)

φ^* and ϖ^*,

$$(C_{\varphi_*} \cdot B)^T = \begin{pmatrix} -4 & 1 & 2 & -1 \\ 2 & 0 & -1 & 0 \\ 1 & -1 & -2 & 2 \\ 3 & -1 & -2 & 0 \\ -2 & 0 & 1 & 0 \\ -1 & 1 & 1 & -1 \\ -1 & 0 & 1 & -1 \\ -1 & 1 & 1 & 0 \\ 0 & 0 & 0 & 0 \\ 0 & 0 & 0 & 0 \end{pmatrix}$$

and

$$(C_{\varpi_*} \cdot B)^T = \begin{pmatrix} -1 & 0 & 0 & 0 \\ 0 & -1 & 0 & -1 \\ 1 & 0 & -1 & 3 \\ -1 & -1 & 0 & 0 \\ 2 & 0 & -1 & 0 \\ -1 & 1 & 1 & -3 \\ -1 & -1 & 0 & 1 \\ 0 & -1 & 0 & 1 \\ 0 & 0 & 0 & 0 \\ 0 & 0 & 0 & 0 \end{pmatrix}.$$

Let $V_{\varphi^*} = \mathrm{Im}(\varphi^*)$ in $H^1(\mathbf{M}, \mathbb{Z})$, and let $V_{\varpi^*} = \mathrm{Im}(\varpi^*)$ in $H^1(\mathbf{M}, \mathbb{Z})$. Since both maps figure in factorizations of the origami map $\pi : \mathbf{M} \to E$, both V_{φ^*} and V_{ϖ^*} will contain V_1, so we set

$$V_2 = (V_{\varphi^*}) \cap V_1^{\perp},$$

and analogously

$$V_3 = (V_{\varpi^*}) \cap V_1^{\perp}.$$

We refrain from presenting the calculations, and merely give bases for V_2 and V_3.

Let v_1^2 and v_2^2 have the coordinate vectors

$$(1, -1, -1, -1, 1, 0, 1, 0)^T \ , \quad \text{and} \quad (5, -2 - 2, -4, 2, 2, 1, 2)^T$$

with respect to $\{s_i^*\}_{i=1}^8$. They span the 2-dimensional symplectic subspace V_2; if $a_1, \dots, a_4$ are the vectors corresponding to the columns of $(C_{\varphi_*} \cdot B)^T$, then $v_1^2 = a_3 - a_2$ and $v_2^2 = a_2 - a_1$.

Let v_1^3 and v_2^3 have the coordinate vectors

$$(1, -1, 2, 1, -2, -2, 2, 1)^T \ , \quad \text{and} \quad (-1, 1, 1, 0, 2, -2, 0, 1)^T$$

with respect to $\{s_i^*\}_{i=1}^8$. They span the 2-dimensional symplectic subspace V_3; if $a'_1, \dots, a'_4$ are the vectors corresponding to the columns of $(C_{\varpi_*} \cdot B)^T$, then $v_1^3 = a'_4 - a'_1$ and $v_2^3 = a'_1 - a'_2$.

The spaces V_2 and V_3 are invariant for the action of the subgroup $\Gamma(2) \leqslant \Gamma_\Theta$. In fact, one can show that only their sum is invariant under Γ_Θ, for one has e. g.

$$\rho(S)(v_1^2) = v_2^3 \quad \text{and} \quad \rho(S)(v_2^2) = -v_1^3 - v_2^3.$$

With respect to the above chosen basis of V_2, the generators of $\Gamma(2)$ act as

$$A_{\rho|V_2(A_1)} = \begin{pmatrix} 1 & 1 \\ 0 & 1 \end{pmatrix} = T \ , \quad A_{\rho|V_2(A_2)} = \begin{pmatrix} 0 & -1 \\ 1 & -1 \end{pmatrix} = ST^{-1} \ , \qquad (9.5)$$

and $-I$ acts as $-I$. We set $\rho_2 = \rho_{|V_2}^{|\Gamma(2)}$. In particular we see that

$$\mathrm{Im}(\rho_2) = \mathrm{Sp}(V_2) \cong \mathrm{SL}_2(\mathbb{Z}). \qquad (9.6)$$

Similarly, with respect to the above chosen basis of V_3, the generators of $\Gamma(2)$ act as

$$A_{\rho|V_3(A_1)} = \begin{pmatrix} 1 & 0 \\ 1 & 1 \end{pmatrix} = ST^{-1}S^{-1} \ , \quad A_{\rho|V_3(A_2)} = \begin{pmatrix} -1 & 1 \\ -1 & 0 \end{pmatrix} = -TS \ , \qquad (9.7)$$

and $-I$ acts as $-I$. We set $\rho_3 = \rho_{|V_3}^{|\Gamma(2)}$. We also see (maybe after conjugation with S^{-1}) that

$$\mathrm{Im}(\rho_3) = \mathrm{Sp}(V_3) \cong \mathrm{SL}_2(\mathbb{Z}). \qquad (9.8)$$

Now let us justify why the splitting of the representation, respectively of the local system carries over to the pVHS. By Proposition 7.15, it follows that if we find a sub-pVHS of the pVHS $(\mathbb{V}, \mathcal{V}^{1,0}, Q)$ coming from the family over the origami curve, then the Q-complementary sub-local system of $\mathbb{V}$ also carries a sub-pVHS. From Theorem 9.3, we get sub-pVHS on the local systems associated with V_1, with V_φ and with V_ϖ. Thus we also get sub-pVHS on V_2 and V_3, and on the symplectic complement of the sum $V_1 + V_2 + V_3$.

Therefore, we dispose of period mappings p_2 and p_3 associated with V_2 and V_3. The computation of their degrees proceeds as follows. First, $\Gamma(2)$ has 3 cusps, whose associated parabolics are e. g.

$$A_1 = \begin{pmatrix} 1 & 2 \\ 0 & 1 \end{pmatrix}, \quad A_2 = \begin{pmatrix} 1 & 0 \\ 2 & 1 \end{pmatrix}, \quad A_3 = -A_1 A_2^{-1} = \begin{pmatrix} 3 & -2 \\ 2 & -1 \end{pmatrix}.$$

To derive the behavior of the period mapping p_2 at the cusps, look at the images of A_1, A_2, A_3. The first two are given in (9.5), and

$$A_{\rho|V_2(A_3)} = -A_{\rho|V_2(A_1)} A^{-1}_{\rho|V_2(A_2)} = \begin{pmatrix} 2 & -1 \\ 1 & 0 \end{pmatrix} = (ST^{-1})T^{-1}(ST^{-1}).$$

Therefore, by Proposition 9.19, p_2 is an unramified 2-to-1 map above the (sole) cusp associated with $i\infty$. Hence, $\deg(p_2) = 2$.

Similarly,

$$A_{\rho|V_3(A_3)} = \begin{pmatrix} 0 & 1 \\ -1 & 2 \end{pmatrix} = -ST^{-2} = (ST^{-1})^{-1}T(ST^{-1}).$$

This and (9.7) together with Proposition 9.19 imply that p_3 is an unramified 2-to-1 map above the cusp associated with $i\infty$.

When we take the symplectic complement of $V_1 + V_2 + V_3$, we obtain a fourth symplectic subspace V_4. Again, we do not carry out the computations, and merely write down a basis for V_4. Let v_1^4, respectively v_2^4 have the coordinate vectors

$$(-3, 0, 3, 1, 0, -1, 2, 2)^T \quad \text{and} \quad (1, -1, -1, 1, -2, 1, -1, 1)^T$$

w. r. t. the basis $\{s_i^*\}_{i=1}^8$, and let $V_4 = \mathrm{span}\{v_1^4, v_1^4\}$.

V_4 is a Γ_Θ-invariant subspace; the action of the generators w. r. t. the chosen basis is given by

$$A_{\rho_{|V_4}(T^2)} = \begin{pmatrix} -1 & -1 \\ 1 & 0 \end{pmatrix} = T^{-1}S\,, \quad A_{\rho_{|V_4}(S)} = \begin{pmatrix} 0 & 1 \\ -1 & 0 \end{pmatrix} = S^{-1}\,, \qquad (9.9)$$
$$A_{\rho_{|V_4}(TSTS^{-1}T^{-1})} = \begin{pmatrix} 1 & 1 \\ 0 & 1 \end{pmatrix} = T.$$

We set $\rho_4 = \rho_{|V_4}$. In particular, by (9.9)

$$\mathrm{Im}(\rho_4) = \mathrm{Sp}(V_4) \cong \mathrm{SL}_2(\mathbb{Z}). \qquad (9.10)$$

Next, consider the period mapping p_4 associated with ρ_4. It follows from (9.9) and Proposition 9.19 that p_4 is an unramified 1-to-1 map above the sole cusp associated with $i\infty$. Therefore, p_4 is an isomorphism.

Putting it all together, one can also show that

$$(V_1 \otimes \mathbb{Q}) \oplus \cdots \oplus (V_4 \otimes \mathbb{Q}) = H^1(\mathbf{M}, \mathbb{Q}),$$

so that we have a complete decomposition over $\mathbb{Q}$ as claimed. ■

From the above computations, it follows that

Corollary 9.26 *The Lyapunov spectrum of* $\mathbf{M}$ *is equal to*

$$1, \tfrac{1}{3}, \tfrac{1}{3}, \tfrac{1}{3}.$$

Note that this is not very surprising as the formula for the sum of the Lyapunov exponents tells us that all 4 must add up to 2 and we get two $\frac{1}{3}$ for free from the covering maps.

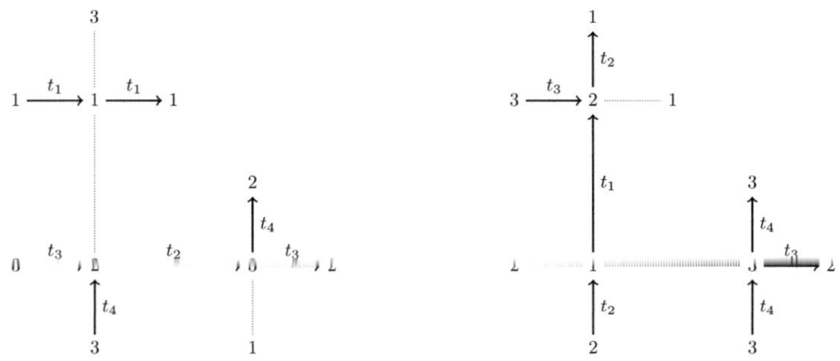

Figure 9.7: *The graphs of the origamis $T \cdot \mathbf{L}_{2,2}$ and $S^{-1}T^{-1} \cdot \mathbf{L}_{2,2}$ with π_1-bases*

The Origami $\mathbf{N}_3$

In this section we resume the discussion of $\mathbf{N}_3$, the first member of the family of Theorem 3.3. First, we will change the numbering of the squares. Let

$$\begin{aligned}\sigma_x &= (1,22,7,19,4,25)(2,3,5,6,8,9)(10,13,16)\cdot\\ &\quad\cdot(11,12,14,15,17,18)(20,21,23,24,26,27)\\ \sigma_y &= (1,18,8,7,15,2)(3,20,19,24,11,10)(4,12,5)\cdot\\ &\quad\cdot(6,26,25,21,14,13)(9,23,22,27,17,16).\end{aligned}$$

Then $\mathbf{N}_3$ is equally well given as the origami with the permutations σ_x and σ_y, as can be checked by comparing Figures 3.1 and 9.8.

Recall that the genus of $\mathbf{N}_3$ is 10, and $\mathbf{N}_3$ lives in the stratum $\Omega\mathcal{M}_{10}(2^9)$. As can be determined e. g. with the help of the origami program, its spin structure is "even". Its affine group is isomorphic to $\mathrm{SL}_2(\mathbb{Z})$, and there is precisely one affine, biholomorphic involution s with derivative $-I$. s has 18 fixed points and $E_s = \mathbf{N}_3 / \langle s \rangle$ has genus 1.

Let us also take a look at the cusps of the origami curve. $\mathbf{N}_3$ decomposes into 4 horizontal respectively vertical cylinders of width 6 and height 1 and one horizontal respectively vertical cylinder of width 3 and height 1.

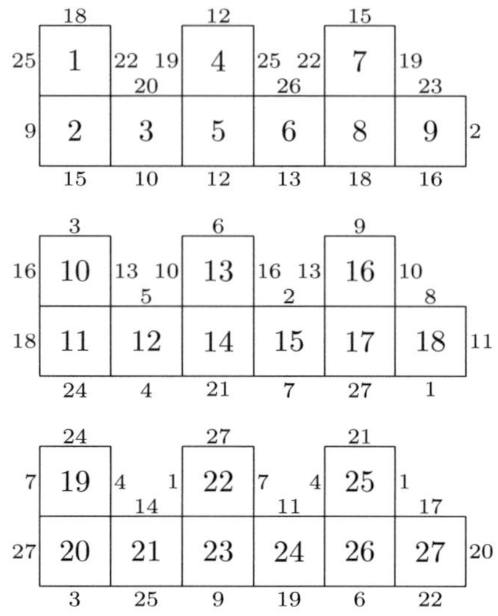

Figure 9.8: *The origami* $\mathbf{N}_3$

From the cylinder decomposition, one deduces that the unique point on $\overline{\mathcal{C}(\mathbf{N}_3)} \cap (\overline{\mathcal{M}_{10}} \setminus \mathcal{M}_{10})$ is a stable curve with two irreducible components of genus 2 and 4 respectively, which meet in 2 points. The genus 2-component has one self-intersection, whereas the genus 4-component has 2.

$\mathbf{N}_3$ covers all the origamis in the $\mathrm{SL}_2(\mathbb{Z})$-orbits of our examples discussed above. In fact, the partition

$$(P_i^1)_{i=1}^9 = (\{1,4,7\},\{2,5,8\},\{3,6,9\},\{10,13,16\}\{11,14,17\},\\ \{12,15,18\},\{19,22,25\},\{20,23,26\},\{21,24,27\}).$$

is acted upon by σ_x and σ_y and induces the covering map $\mathbf{N}_3 \to \mathbf{M}$: Send the square i to the square j such that $j \in P_i^1$. Analogously, one constructs the covering map $\mathbf{N}_3 \to T \cdot \mathbf{M}$ from

$$(P_i^2)_{i=1}^9 = (\{1,8,15\},\{2,7,18\},\{3,11,19\},\{4,5,12\}\{6,14,25\},\\ \{9,17,22\},\{10,20,24\},\{13,21,26\},\{16,23,27\}),$$

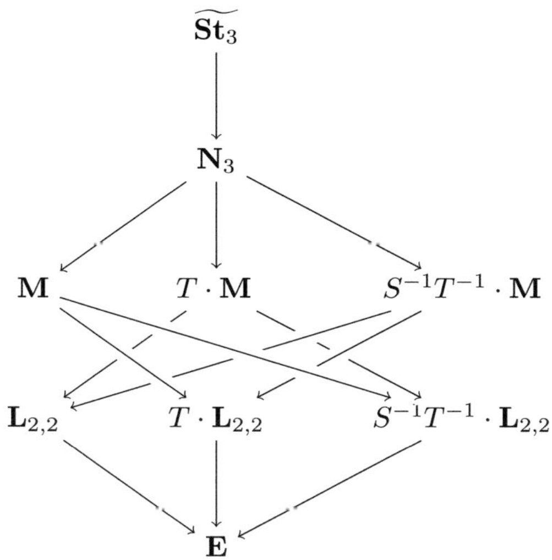

Figure 9.9: *The poset of intermediate covers of* $\mathbf{N}_3$

and the covering map $\mathbf{N}_3 \to S^{-1}T^{-1} \cdot \mathbf{M}$ from

$$(P_i^3)_{i=1}^9 = (\{1,13,19\},\{2,14,20\},\{3,15,21\},\{4,16,22\}\{5,17,23\},\\ \{6,18,24\},\{7,10,25\},\{8,11,26\},\{9,12,27\}).$$

The poset of intermediate covers for $\mathbf{N}_3$ is shown in Figure 9.9. Since $\mathbf{N}_3$ covers many smaller origamis, we are bound to see pullbacks of representations from lower genus in the monodromy representation of $\mathbf{N}_3$. In fact, we have

Theorem 9.27 *The representation* $\rho : \mathrm{Aff}(\mathbf{N}_3) \to \mathrm{Sp}(H^1(\mathbf{N}_3, \mathbb{Q}))$ *restricted to* $\Gamma(2)$ *splits over* $\mathbb{Q}$ *into nine symplectic subrepresentations* (ρ_i, U_i)*,* $i = 1, \dots, 9$.

Besides the trivial representation (ρ_1, U_1)*, there are*

- *three 2-dimensional representations* (ρ_i, U_i)*,* $i = 2, 3, 4$*, which are* *pullbacks of the non-trivial representations of* $\mathbf{L}_{2,2}$*,* $T \cdot \mathbf{L}_{2,2}$ *and* $S^{-1}T^{-1} \cdot \mathbf{L}_{2,2}$ *respectively.*

- *three 2-dimensional representations* (ρ_i, U_i)*,* $i = 5, 6, 7$*, which are pullbacks of the representation* ρ_4 *on* $\mathbf{M}$*,* $T \cdot \mathbf{M}$ *and* $S^{-1}T^{-1} \cdot \mathbf{M}$ *respectively.*
- *one 2-dimensional representation* (ρ_8, U_8)*, which is precisely the* 1*-eigenspace of* s *and on which* $\Gamma(2)$ *acts by a cyclic group of order* 3
- *one 4-dimensional representation* (ρ_9, U_9) *such that the action of* $\Gamma(2)$ *on* U_9 *is by a finite group of order* 24*, which is isomorphic to* $\mathrm{SL}_2(\mathbb{Z}/(3))$.

Note that from Remark 8.18 and the above, it follows that $\mathbf{N}_3$ has a fixed part in its family of Jacobians of complex dimension 3.

As a direct consequence, we obtain the Lyapunov spectrum of $\mathbf{N}_3$. Note that the sum of the Lyapunov exponents equals 3 in accordance with the formula of Proposition 8.13.

Corollary 9.28 *The Lyapunov spectrum of* $\mathbf{N}_3$ *is given by*

$$1, \underbrace{\tfrac{1}{3}, \ldots, \tfrac{1}{3}}_{6}, \underbrace{0, \ldots, 0}_{3}.$$

The computations needed to justify Theorem 9.27 proceed as in the above examples with the drawback that the dimensions are much higher (28 for the initial basis and 20 for the basis of the homology of the closed surface). So we refrain from presenting them here.

9.5 Equivalence of Rank 2-Period Mappings

Following the theme of Theorem 9.1, it would be interesting to know which isomorphism classes of irreducible sub-pVHS can occur in the pVHS of a family coming from a Teichmüller curve. Of course, Proposition 7.28 provides a partial answer. There is always one sub-local system $\mathbb{L}_1$, defined over the trace field of the Teichmüller curve, carrying a pVHS, whose Kodaira-Spencer map is maximal Higgs. Then there are its Galois conjugates, and there is a remainder $\mathbb{M}$, defined over $\mathbb{Q}$ and also carrying a pVHS.

In the following, we set up a toy model, the set of rank 2-pVHS defined over $\mathbb{Q}$, and we define an ordering and an equivalence relation on this set. Using the computations in Section 9.4, we give an example of a pVHS of rank 2, defined over $\mathbb{Q}$, which does not come from genus 2.

If $g \in G$, let $c_g : G \to G$, $h \mapsto ghg^{-1}$ denote the inner automorphism of G obtained by conjugation with g. If $H \leqslant G$ and $g \in G$, then H^g is the subgroup $c_g(H) = gHg^{-1} \leqslant G$.

Definition 9.29 A *period datum* (in rank 2, over $\mathbb{Q}$) is a triple $P = (p, \Gamma, \rho)$ such that $p : \mathbb{H} \to \mathbb{H}$ is a holomorphic map, $\Gamma \leqslant \mathrm{SL}_2(\mathbb{R})$ is a cofinite Fuchsian group and $\rho : \Gamma \to \mathrm{SL}_2(\mathbb{Z})$ is a group homomorphism whose image has finite index, and p is ρ-equivariant. We denote the induced map $\Gamma \backslash \mathbb{H} \to \rho(\Gamma) \backslash \mathbb{H}$ by $\iota(p)$.

Remark 9.30 Assuming that in $P = (p, \Gamma, \rho)$, the Fuchsian group is torsion-free, we can define a pVHS of rank 2, defined over $\mathbb{Q}$ on the curve $\mathcal{C} = \Gamma \backslash \mathbb{H}$. Take a base point $c \in \mathcal{C}$, and consider the local system associated with $\rho : \Gamma = \pi_1(\mathcal{C}, c) \to \mathrm{Sp}(2, \mathbb{Z}) = \mathrm{SL}_2(\mathbb{Z})$. To define a pVHS on $\Gamma \backslash \mathbb{H}$, take the pullback of the universal bundle over the period domain $\mathbb{H}$, pull it back via p and push it down to $\Gamma \backslash \mathbb{H}$.

The following definition is motivated by two observations: We would like to allow a change of the base point of the fundamental group, and we would like to allow a change of the symplectic basis of the local system.

Definition 9.31 Let $P = (p, \Gamma, \rho)$ and $Q = (q, \Delta, \sigma)$ be period data.

a) We say that P is equivalent to Q, short $P \sim Q$, if there exist $\alpha \in \mathrm{SL}_2(\mathbb{R})$, $\beta \in \mathrm{SL}_2(\mathbb{Z})$ such that
   - $q = \beta \circ p \circ \alpha^{-1}$,
   - $\Gamma^\alpha = \Delta$ and $\rho(\Gamma)^\beta = \sigma(\Delta)$,
   - $\sigma = c_\beta \circ \rho \circ c_{\alpha^{-1}}$.

b) We say that Q dominates P, short $Q \succcurlyeq P$, if there exist elements $\alpha \in \mathrm{SL}_2(\mathbb{R})$ and $\beta \in \mathrm{SL}_2(\mathbb{Z})$ such that $\Gamma^\alpha \leqslant \Delta$, $\rho(\Gamma)^\beta \leqslant \sigma(\Delta)$ and the diagram

$$\begin{array}{ccc} \Gamma^\alpha \backslash \mathbb{H} & \xrightarrow{\iota(\beta \circ p \circ \alpha^{-1})} & \rho(\Gamma)^\beta \backslash \mathbb{H} \\ \downarrow & & \downarrow \\ \Delta \backslash \mathbb{H} & \xrightarrow[\iota(q)]{} & \sigma(\Delta) \backslash \mathbb{H} \end{array} \tag{9.11}$$

commutes. (The vertical arrows are induced by the inclusions.)

Lemma 9.32 *Let $P = (p, \Gamma, \rho)$ and $Q = (q, \Delta, \sigma)$ be period data.*

a) $\sim$ is an equivalence relation.

b) For every $\gamma \in \rho(\Gamma)$, we have $P \sim P^\gamma = (\gamma \circ p, \Gamma, c_\gamma \circ \rho)$.

c) If $P \succcurlyeq Q$ and $Q \succcurlyeq P$, then $P \sim Q$.

d) $\succcurlyeq$ is reflexive and transitive.

Proof: a) follows directly from the definition.

b) P^γ is a well-defined period datum, as for all $z \in \mathbb{H}$ and all $\tilde{\gamma} \in \Gamma$, we have

$$\gamma p(\tilde{\gamma}(z)) = \gamma \rho(\tilde{\gamma}) p(z) = (c_\gamma \circ \rho)(\tilde{\gamma}) \gamma p(z).$$

Letting $\beta = \gamma$, and $\alpha = \mathrm{id}$ in the definition of $\sim$, we obviously have $P \sim P^\gamma$.

c) Since $Q \succcurlyeq P$, there are $\alpha \in \mathrm{SL}_2(\mathbb{R})$, $\beta \in \mathrm{SL}_2(\mathbb{Z})$ such that

$$\Gamma^\alpha \subset \Delta \quad \text{and} \quad \rho(\Gamma)^\beta \subset \sigma(\Delta),$$

and such that we have a commutative diagram as above. From $P \succcurlyeq Q$, we get $\alpha' \in \mathrm{SL}_2(\mathbb{R})$, $\beta' \in \mathrm{SL}_2(\mathbb{Z})$ such that

$$\Delta^{\alpha'} \subset \Gamma \quad \text{and} \quad \sigma(\Delta)^{\beta'} \subset \rho(\Gamma),$$

and another commutative diagram. Since $\Gamma^{\alpha\alpha'} \subset \Delta^{\alpha'} \subset \Gamma$, we see that Γ and Δ are conjugate, for $\Gamma^{\alpha\alpha'} = \Gamma$, as $\Gamma^{\alpha\alpha'}$ is a subgroup of Γ with the same volume. The same holds for $\rho(\Gamma)$ and $\sigma(\Delta)$. Therefore, the vertical arrows in the commutative diagram (9.11) are isomorphisms. The maps q and $\beta p\alpha^{-1}$ are both lifts of $\tau : \mathbb{H} \mapsto \sigma(\Delta)\backslash\mathbb{H}$, where τ is the composition of $\mathbb{H} \to \Gamma^\alpha\backslash\mathbb{H}$, followed by either way through the diagram from the upper left to the lower right corner. Hence q and $\beta p\alpha^{-1}$ differ only by an element γ of the deck group $\sigma(\Delta)$, i. e. $\gamma g = \beta p\alpha^{-1}$. We show that $Q^\gamma \sim P$.

Let $\delta \in \Gamma^\alpha$. Then for all $z \subset \mathbb{H}$

$$\begin{aligned}\beta p\alpha^{-1}(\delta\cdot z) &= \beta p(c_{\alpha^{-1}}(\delta)\alpha^{-1}\cdot z)\\ &= \beta\rho(c_{\alpha^{-1}}(\delta))p(\alpha^{-1}\cdot z)\\ &= c_\beta\circ\rho\circ c_{\alpha^{-1}}(\delta)\ \beta p\alpha^{-1}(z).\end{aligned}$$

and

$$\begin{aligned}\beta p\alpha^{-1}(\delta\cdot z) &= \gamma q(\delta\cdot z)\\ &= \gamma\sigma(\delta)q(z)\\ &= c_\gamma\circ\sigma(\delta)\gamma q(z)\\ &= c_\gamma\circ\sigma(\delta)\ \beta p\alpha^{-1}(z).\end{aligned}$$

Since this holds for all $z \in \mathbb{H}$, and since p is not constant, we conclude that

$$c_\gamma\circ\sigma(\delta) = c_\beta\circ\rho\circ c_{\alpha^{-1}}(\delta)$$

for all $\delta \in \Gamma^\alpha$. Therefore $P \sim Q^\gamma \sim Q$.

d) is straightforward and will be omitted. ■

For the following definition, recall from Proposition 7.28 that for a pVHS on coming from a Veech surface (X,ω) in genus 2, either $(K(X,\omega) : \mathbb{Q}) = 2$, and we have two Galois conjugate local systems on (a finite cover of) the Teichmüller curve, or $\mathbb{K}(X,\omega) = \mathbb{Q}$. Then (X,ω) is arithmetic by [GJ00]. If (X,ω) is arithmetic, then we find an origami on the Teichmüller curve of (X,ω). This motivates the following

Definition 9.33 A period datum $P = (p,\Gamma,\rho)$ *comes from* $\Omega\mathcal{M}_2$, if it is dominated by $Q = (q,\Delta,\sigma)$, where $\Delta = \Gamma(\mathbf{O})$ with $\mathbf{O}$ an origami in genus 2, σ is the non-trivial sub-representation of $\mathrm{Aff}(\mathbf{O}) \to \mathrm{Sp}(H^1(\mathbf{O},\mathbb{Z}))$, and

q is the period mapping associated with the sub-pVHS on the local system corresponding to σ.

Proposition 9.34 *The subrepresentation ρ_4 of the origami* $\mathbf{M}$ *does not come from $\Omega\mathcal{M}_2$.*

PROOF: Assume that $P = (p_4, \Gamma_\Theta, \rho_4)$ comes from $\Omega\mathcal{M}_2$, and let $Q = (q, \Delta, \sigma)$ be a period datum coming from an origami $\mathbf{O}$ of genus 2 that dominates it. From (9.11), it follows that $\mathbf{O}$ is in the stratum $\Omega\mathcal{M}_2(2)$. For otherwise, the non-trivial Lyapunov exponent of $\mathbf{O}$ is $1/2$ by Proposition 8.15. Hence by Theorem 9.18

$$\frac{1}{2} = \frac{\deg(\iota(q))\,\mathrm{vol}(\sigma(\Delta)\backslash\mathbb{H})}{\mathrm{vol}(\Delta\backslash\mathbb{H})}.$$

But

$$\frac{1}{3} = \frac{\deg(p_4)\,\mathrm{vol}(\mathrm{Im}(\rho_4)\backslash\mathbb{H})}{\mathrm{vol}(\Gamma_\Theta\backslash\mathbb{H})} = \frac{\deg(\iota(q)\,\mathrm{vol}(\sigma(\Delta)\backslash\mathbb{H})}{\mathrm{vol}(\Delta\backslash\mathbb{H}})$$

by the commutativity of (9.11). So $\mathbf{O}$ is in the stratum $\Omega\mathcal{M}_2(2)$. By assumption, there is $\alpha \in \mathrm{SL}_2(\mathbb{R})$, and $\beta \in \mathrm{SL}_2(\mathbb{Z})$ such that $\Gamma_\Theta^\alpha \leqslant \Delta$ and $\rho_4(\Gamma_\Theta)^\beta \leqslant \sigma(\Delta)$. On the other hand, we know from [EMS03] and [LR06] what the index of Δ in $\mathrm{SL}_2(\mathbb{Z})$ is: If the number of squares n of $\mathbf{O}$ is 3, then $(\mathrm{SL}_2(\mathbb{Z}) : \Delta) = 3$. Otherwise, $(\mathrm{SL}_2(\mathbb{Z}) : \Delta)$ is given by

$$\frac{3}{8}(n-2)n^2 \prod_{p|n}(1 - \frac{1}{p^2})$$

if n is even, and $n \geq 4$, and it is given either by

$$\frac{3}{16}(n-1)n^2 \prod_{p|n}(1 - \frac{1}{p^2})$$

or by

$$\frac{3}{16}(n-3)n^2 \prod_{p|n}(1 - \frac{1}{p^2})$$

if n is odd and $n \geq 5$ – depending on the spin invariant of $\mathbf{O}$. Note that there are no origamis in $\Omega\mathcal{M}_2(2)$ with less than 3 squares. By inspecting the above formulas, one finds that for $n > 3$, the index of Δ in $\mathrm{SL}_2(\mathbb{Z})$ would be ≥ 4, which is impossible, since Γ_Θ^α has index 3 and is contained in Δ. It follows that $\Delta = \Gamma_\Theta^\alpha$, and in particular

$$\Delta \in \{\Gamma(\mathbf{L}_{2,2}), \Gamma(T \cdot \mathbf{L}_{2,2}), \Gamma(S^{-1}T^{-1} \cdot \mathbf{L}_{2,2})\}.$$

Therefore, $\sigma = \rho_2^{\mathbf{L}_{2,2}} \circ c_{\alpha^{-1}}$ with $\rho_2^{\mathbf{L}_{2,2}} = \rho_2$ from Remark 9.24. Now $c_\alpha(T^2)$ is a parabolic element of Γ_Θ^α which is mapped on the one hand to

$$\sigma(c_\alpha(T^2)) = \rho_2^{\mathbf{L}_{2,2}}(T^2)$$

and on the other hand to

$$c_\beta \circ \rho_4 \circ c_{\alpha^{-1}}(c_\alpha(T^2)) = c_\beta \circ \rho_4(T^2).$$

In particular, $\rho_2^{\mathbf{L}_{2,2}}(T^2)$ is a parabolic element, conjugate to $\left(\begin{smallmatrix} 1 & 1 \\ 0 & 1 \end{smallmatrix}\right)$, and $c_\beta \circ \rho_4(T^2)$ is an elliptic element, conjugate to $\left(\begin{smallmatrix} -1 & -1 \\ 1 & 0 \end{smallmatrix}\right)$. By assumption, the diagram (9.11) commutes. Since $\mathrm{Im}(\rho_4)$ and $\mathrm{Im}(\rho_2^{\mathbf{L}_{2,2}})$ are both equal to $\mathrm{SL}_2(\mathbb{Z})$, the vertical arrows are isomorphisms. By Lemma 9.20, $\iota(q)$ maps the cusp corresponding to the fixed point of $c_\alpha(T^2)$ to a cusp. In order to obtain a contradiction to the commutativity of (9.11), we invoke the nilpotent orbit theorem [Sch73, Theorem 4.9, Corollary 4.11]: As $c_\beta \circ \rho_4(T^2)$ has finite order, $\iota(\beta \circ p_4 \circ \alpha^{-1})$ maps the cusps corresponding to the fixed point of $c_\alpha(T^2)$ to an interior point of $\rho_4(\Gamma_\Theta)\backslash \mathbb{H}$. ■

One may wish to refine the notion of dominance in the following way.

Definition 9.35 Two period data (p, Γ, ρ) and (q, Δ, σ) are called *commensurable*, if there is a finite-index subgroup Γ' of Γ and a finite index subgroup Δ' of Δ such that

$$(p, \Gamma', \rho_{|\Gamma'}) \sim (q, \Delta', \sigma_{|\Delta'}).$$

Then it is no longer clear that $P = (p_4, \Gamma_\Theta, \rho_4)$ does not come from genus 2 in the sense that it is not commensurable to any period datum in genus 2.

A Appendix

The table on the next page is the character table of the group K_3. The first row contains a representative of each of the 15 conjugacy classes of elements in K_3. The second and third row contain the number of elements conjugate to the representative and the order of the representative.

Conj.cl.	$[1,1,1]$	$[\tau,1,1]$	$[1,\tau,1]$	$[1,1,\tau]$	$[\tau,\tau,1]$	$[\tau,1,\tau]$	$[1,\tau,\tau]$	$[\tau,\tau,\tau]$	$[\tau^2,\tau,\tau]$	$[\sigma,\sigma,1]$	$[1,\sigma,\sigma]$	$[\sigma,1,\sigma]$	$[\sigma,\sigma,\tau]$	$[\sigma,\tau,\sigma]$	$[\tau,\sigma,\sigma]$
# Elem.	1	2	2	2	4	4	4	4	4	9	9	9	18	18	18
Order	1	3	3	3	3	3	3	3	3	2	2	2	6	6	6
χ_1	1	1	1	1	1	1	1	1	1	1	1	1	1	1	1
χ_2	1	1	1	1	1	1	1	1	1	-1	1	-1	-1	-1	1
χ_3	1	1	1	1	1	1	1	1	1	1	-1	-1	1	-1	-1
χ_4	1	1	1	1	1	1	1	1	1	-1	-1	1	-1	1	-1
χ_5	2	2	-1	2	-1	2	-1	-1	-1	0	0	-2	0	1	0
χ_6	2	2	-1	2	-1	2	-1	-1	-1	0	0	2	0	-1	0
χ_7	2	-1	2	2	-1	-1	2	-1	-1	0	-2	0	0	0	1
χ_8	2	-1	2	2	-1	-1	2	-1	-1	0	2	0	0	0	-1
χ_9	2	2	2	-1	2	-1	-1	-1	-1	-2	0	0	1	0	0
χ_{10}	2	2	2	-1	2	-1	-1	-1	-1	2	0	0	-1	0	0
χ_{11}	4	-2	-2	4	1	-2	-2	1	1	0	0	0	0	0	0
χ_{12}	4	4	-2	-2	-2	-2	1	1	1	0	0	0	0	0	0
χ_{13}	4	-2	4	-2	-2	1	-2	1	1	0	0	0	0	0	0
χ_{14}	4	-2	-2	-2	1	1	1	$-2-3\zeta_3$	$1+3\zeta_3$	0	0	0	0	0	0
χ_{15}	4	-2	-2	-2	1	1	1	$1+3\zeta_3$	$-2-3\zeta_3$	0	0	0	0	0	0

Table A.1: *Character table of* K_3

Bibliography

[AV07] Artur Avila and Marcelo Viana, *Simplicity of Lyapunov spectra: proof of the Zorich-Kontsevich conjecture*, Acta Math. **198** (2007), no. 1, 1–56. MR 2316268 (2008m:37010)

[Bai07] Matt Bainbridge, *Euler characteristics of Teichmüller curves in genus two*, Geom. Topol. **11** (2007), 1887–2073. MR 2350471 (2009c:32025)

[Bau09] Oliver Bauer, *Familien von Jacobivarietäten über Origamikurven*, Ph.D. thesis, Karlsruhe, 2009.

[BL04] Christina Birkenhake and Herbert Lange, *Complex abelian varieties*, second ed., Grundlehren der Mathematischen Wissenschaften [Fundamental Principles of Mathematical Sciences], vol. 302, Springer-Verlag, Berlin, 2004.

[BM10a] Irene Bouw and Martin Möller, *Differential equations associated with nonarithmetic Fuchsian groups*, J. Lond. Math. Soc. (2) **81** (2010), no. 1, 65–90. MR 2580454

[BM10b] ______, *Teichmüller curves, triangle groups, and Lyapunov exponents*, Annals of Math. **172** (2010), no. 1, 139–185.

[CFS82] Isaak P. Cornfeld, Sergej V. Fomin, and Jakov G. Sinaj, *Ergodic theory*, Die Grundlehren der mathematischen Wissenschaften in Einzeldarstellungen ; 245, Springer, New York, 1982.

[CM11] Dawei Chen and Martin Möller, *Non-varying sums of Lyapunov exponents for abelian differentials in low genus*, preprint, 2011.

[Del70] Pierre Deligne, *Equations différentielles à points singuliers réguliers*, Lecture notes in mathematics ; 163, Springer, Berlin, 1970.

[Del71a] ______, *Théorie de Hodge. II*, Inst. Hautes Études Sci. Publ. Math. (1971), no. 40, 5–57.

[Del71b] ______, *Travaux de Griffiths*, Séminaire Bourbaki vol. 1969/70 Exposés 364–381, Lecture Notes in Mathematics, vol. 180, Springer Berlin / Heidelberg, 1971, pp. 213–237.

[Del87] ______, *Un théorème de finitude pour la monodromie*, Discrete groups in geometry and analysis (New Haven, Conn., 1984), Progr. Math., vol. 67, Birkhäuser Boston, Boston, MA, 1987, pp. 1–19.

[DM69] P. Deligne and D. Mumford, *The irreducibility of the space of curves of given genus*, Inst. Hautes Études Sci. Publ. Math. (1969), no. 36, 75–109. MR 0262240 (41 #6850)

[EE69] Clifford J. Earle and James Eells, *A fibre bundle description of Teichmüller theory*, J. Differential Geometry **3** (1969), 19–43. MR 0276999 (43 #2737a)

[EG97] Clifford J. Earle and Frederick P. Gardiner, *Teichmüller disks and Veech's $\mathcal{F}$-structures*, Extremal Riemann surfaces (San Francisco, CA, 1995), Contemp. Math., vol. 201, Amer. Math. Soc., Providence, RI, 1997, pp. 165–189. MR 1429199 (97k:32031)

[EKZ10a] A. Eskin, M. Kontsevich, and A. Zorich, *Lyapunov spectrum of square-tiled cyclic covers*, preprint, 2010.

[EKZ10b] ______, *Sum of Lyapunov exponents of the Hodge bundle with respect to the Teichmüller geodesic flow*, preprint, 2010.

[EMS03] Alex Eskin, Howard Masur, and Martin Schmoll, *Billiards in rectangles with barriers*, Duke Math. J. **118** (2003), no. 3, 427–463.

[EO01] Alex Eskin and Andrei Okounkov, *Asymptotics of numbers of branched coverings of a torus and volumes of moduli spaces of holomorphic differentials*, Invent. Math. **145** (2001), no. 1, 59–103. MR 1839286 (2002g:32018)

[FMZ10] G. Forni, C. Matheus, and A. Zorich, *Square-tiled cyclic covers*, preprint, 2010.

[For81] Otto Forster, *Lectures on Riemann Surfaces*, Springer-Verlag, New York, 1981.

[For02] Giovanni Forni, *Deviation of ergodic averages for area-preserving flows on surfaces of higher genus*, Ann. of Math. (2) **155** (2002), no. 1, 1–103. MR 1888794 (2003g:37009)

[For06] ______, *On the Lyapunov exponents of the Kontsevich-Zorich cocycle*, Handbook of dynamical systems. Vol. 1B, Elsevier B. V., Amsterdam, 2006, pp. 549–580. MR 2186248 (2010c:37004)

[GJ00] Eugene Gutkin and Chris Judge, *Affine mappings of translation surfaces: geometry and arithmetic*, Duke Math. J. **103** (2000), no. 2, 191–213. MR 1760625 (2001h:37071)

[Gri68a] Phillip A. Griffiths, *Periods of integrals on algebraic manifolds. I. Construction and properties of the modular varieties*, Amer. J. Math. **90** (1968), 568–626.

[Gri68b] ______, *Periods of integrals on algebraic manifolds. II. Local study of the period mapping*, Amer. J. Math. **90** (1968), 805–865.

[Gri70] ______, *Periods of integrals on algebraic manifolds: Summary of main results and discussion of open problems*, Bull. Amer. Math. Soc. **76** (1970), 228–296. MR 0258824 (41 #3470)

[Gro62a] A. Grothendieck, *Techniques de construction en géométrie analytique. I: Description axiomatique de l'espace de Teichmüller et de ses variantes*, Familles d'Espaces Complexes et Fondements de la Geom. Anal., Sem. H. Cartan 13 (1960/61), No.7-8, 33 p., 1962.

[Gro62b] ______, *Techniques de construction en géométrie analytique. VI: Étude locale des morphismes; germes d'espaces analytiques, platitude, morphismes simples*, Familles d'Espaces Complexes et Fondements de la Geom. Anal., Sem. H. Cartan 13 (1960/61), No.13, 13 p., 1962.

[Gro62c] ______, *Techniques de construction en géométrie analytique. X: Construction de l'espace de Teichmüller.*, Familles d'Espaces Complexes et Fondements de la Geom. Anal., Sem. H. Cartan 13 (1960/61), No.17, 20 p. (1962)., 1962.

[Har04] Robin Hartshorne, *Algebraic geometry*, 13.[print.] ed., Graduate texts in mathematics ; 52, Springer, New York, [2004].

[Hat02] Allen Hatcher, *Algebraic Topology*, Cambridge University Press, 2002.

[Her06] Frank Herrlich, *Teichmüller Curves Defined by Characteristic Origamis*, Contemporary Mathematics **397** (2006), 133–144.

[HL06] Pascal Hubert and Samuel Lelièvre, *Prime Arithmetic Teichmüller Disks in $\mathcal{H}(2)$*, Israel Journal of Mathematics **151** (2006), 281–321.

[HM98] Joe Harris and Ian Morrison, *Moduli of curves*, Graduate texts in mathematics ; 187, Springer, New York, 1998.

[Hoo09] W. Patrick Hooper, *Grid graphs and lattice surfaces*, preprint, 2009.

[HS06] Frank Herrlich and Gabriela Schmithüsen, *On the boundary of Teichmüller disks in Teichmüller space and in Schottky space*, Handbook of Teichmüller theory (2006).

[HS08] ———, *An extraordinary origami curve*, Math. Nachrichten **281** (2008), no. 2, 219–237.

[Hub06] John H. Hubbard, *Teichmüller theory and applications to geometry, topology, and dynamics*, vol. 1: Teichmüller theory, Ithaca, Matrix editions, 2006.

[Kan03] Ernst Kani, *Hurwitz spaces of genus 2 covers of an elliptic curve*, Collect. Math. **54** (2003), no. 1, 1–51.

[Kan06] ———, *The number of genus 2 covers of an elliptic curve*, Manuscripta Math. **121** (2006), no. 1, 51–80.

[Kat92] Svetlana Katok, *Fuchsian groups*, Chicago lectures in mathematics series, Univ. of Chicago Press, Chicago, Ill. [u.a.], 1992.

[Kon97] M. Kontsevich, *Lyapunov exponents and Hodge theory*, The mathematical beauty of physics (Saclay, 1996), Adv. Ser. Math. Phys., vol. 24, World Sci. Publ., River Edge, NJ, 1997, pp. 318–332. MR 1490861 (99b:58147)

[Kre10] Karsten Kremer, *Invariants of complex and p-adic origami-curves*, Ph.D. thesis, Karlsruhe, 2010.

[KZ03] Maxim Kontsevich and Anton Zorich, *Connected components of the moduli spaces of Abelian differentials with prescribed singularities*, Invent. Math. **153** (2003), no. 3, 631–678. MR 2000471 (2005b:32030)

[Loc05] Pierre Lochak, *On Arithmetic Curves in the Moduli Space of Curves*, Journal of the Institute of Mathematics of Jussieu **4** (2005), no. 3, 443–508.

[Loo94] Eduard Looijenga, *Smooth Deligne-Mumford compactifications by means of Prym level structures*, J. Algebraic Geom. **3** (1994), no. 2, 283–293. MR 1257324 (94m:14029)

[LR06] Samuel Lelièvre and Emmanuel Royer, *Orbitwise countings in $\mathcal{H}(2)$ and quasimodular forms*, Int. Math. Res. Not. (2006), Art. ID 42151, 30. MR 2233718 (2007h:11049)

[Mas75] Howard Masur, *On a class of geodesics in Teichmüller space*, Ann. of Math. (2) **102** (1975), no. 2, 205–221. MR 0385173 (52 #6038)

[Mas82] ———, *Interval exchange transformations and measured foliations*, Ann. of Math. (2) **115** (1982), no. 1, 169–200. MR 644018 (83e:28012)

[Mas09] ———, *Geometry of Teichmüller space with the Teichmüller metric*, Surveys in differential geometry. Vol. XIV. Geometry of Riemann surfaces and their moduli spaces, Surv. Differ. Geom., vol. 14, Int. Press, Somerville, MA, 2009, pp. 295–313. MR 2655331

[McM03] Curtis T. McMullen, *Billiards and Teichmüller curves on Hilbert modular surfaces*, Journal of the American Mathematical Society **4** (2003), no. 16, 857–885.

[McM06] ———, *Prym varieties and Teichmüller curves*, Duke Math. J. **133** (2006), no. 3, 569–590. MR 2228463 (2007a:32018)

[Möl05a] Martin Möller, *Shimura and Teichmüller curves*, to appear in Journal Modern Dynamics, 2005.

[Möl05b] ———, *Teichmüller Curves, Galois Actions and $\widehat{GT}$-Relations*, Math. Nachrichten **278** (2005), no. 9, 1061–1077.

[Möl06] ———, *Variations of Hodge structures of a Teichmüller curve*, J. Amer. Math. Soc. **19** (2006), no. 2, 327–344.

[PS03] C. A. M. Peters and J. H. M. Steenbrink, *Monodromy of variations of Hodge structure*, Acta Appl. Math. **75** (2003), no. 1-3, 183–194, Monodromy and differential equations (Moscow, 2001). MR 1975567 (2004c:14014)

[Rue79] David Ruelle, *Ergodic theory of differentiable dynamical systems*, Inst. Hautes Études Sci. Publ. Math. (1979), no. 50, 27–58. MR 556581 (81f:58031)

[Sch73] Wilfried Schmid, *Variation of Hodge structure: the singularities of the period mapping*, Invent. Math. **22** (1973), 211–319.

[Sch04] Gabriela Schmithüsen, *An Algorithm for Finding the Veech Group of an Origami*, Experimental Mathematics **13** (2004), no. 4, 459–472.

[Sch05a] Gabriela Schmithüsen, *Veech groups of origamis*, PhD thesis, Universität Karlsruhe, Fakultät für Mathematik, 2005.

[Sch05b] Martin Schmoll, *Spaces of elliptic differentials*, Algebraic and topological dynamics, Contemp. Math., vol. 385, Amer. Math. Soc., Providence, RI, 2005, pp. 303–320. MR 2180242 (2007e:14044)

[Sch06] Gabriela Schmithüsen, *Examples for Veech groups of origamis*, The geometry of Riemann surfaces and abelian varieties, Contemp. Math., vol. 397, Amer. Math. Soc., Providence, RI, 2006, pp. 193–206.

[Ser96] Jean Pierre Serre, *Linear representations of finite groups*, corr. 5. print. ed., Graduate texts in mathematics ; 42, Springer, New York, NY, 1996, Literaturangaben; (Berlin ...) Pp. ; (New York ...) Pp.

[Sti80] John Stillwell, *Classical topology and combinatorial group theory*, Graduate texts in mathematics ; 72, Springer, New York, 1980.

[SW04] John Smillie and Barak Weiss, *Minimal sets for flows on moduli space*, Israel J. Math. **142** (2004), 249–260. MR 2085718 (2005g:37067)

[Vee82] William A. Veech, *Gauss measures for transformations on the space of interval exchange maps*, Ann. of Math. (2) **115** (1982), no. 1, 201–242. MR 644019 (83g:28036b)

[Vee86] ______, *The Teichmüller geodesic flow*, Ann. of Math. (2) **124** (1986), no. 3, 441–530. MR 866707 (88g:58153)

[Vee89] ______, *Teichmüller curves in moduli space, Eisenstein series and an application to triangular billiards*, Inventiones Mathematicae **97** (1989), no. 3, 553–583.

[Vee90] ______, *Moduli spaces of quadratic differentials*, J. Analyse Math. **55** (1990), 117–171. MR 1094714 (92e:32014)

[Voi02] Claire Voisin, *Hodge theory and complex algebraic geometry*, Cambridge studies in advanced mathematics ; 76, vol. 1: 1. publ, Cambridge University Press, Cambridge, 2002.

[Voi03] ______, *Hodge theory and complex algebraic geometry*, Cambridge studies in advanced mathematics ; 77, vol. 2: 1. publ, Cambridge University Press, Cambridge, 2003.

[Vor96] Ya. B. Vorobets, *Plane structures and billiards in rational polygons: the Veech alternative*, Uspekhi Mat. Nauk **51** (1996), no. 5(311), 3–42. MR 1436653 (97j:58092)

[VZ04] Eckart Viehweg and Kang Zuo, *A characterization of certain Shimura curves in the moduli stack of abelian varieties*, J. Differential Geom. **66** (2004), no. 2, 233–287.

[Wal82] Peter Walters, *An introduction to ergodic theory*, Graduate texts in mathematics ; 79, Springer, New York, 1982.

[Wei35] André Weil, *Über Matrizenringe auf Riemannschen Flächen und den Riemann-Rochschen Satz*, Abhandlungen aus dem Mathematischen Seminar der Universität Hamburg **11** (1935), 110–115.

[Wri11] Alex Wright, *Abelian square-tiled surfaces*, preprint, 2011.

[Zor96] Anton Zorich, *Finite Gauss measure on the space of interval exchange transformations. Lyapunov exponents*, Ann. Inst. Fourier (Grenoble) **46** (1996), no. 2, 325–370. MR 1393518 (97f:58081)

[Zor06] ———, *Flat Surfaces*, Frontiers in Number Theory, Physics and Geometry **1** (2006), 439–586.

Index

affine, 9
affine group, 9
algebraic primitivity, 12
arithmetic, 13

characteristic, 23
cocyle, 71
commensurable, 125
complete, 71
complex variation, 58

Deligne extension, 66

ergodic, 71

factor of automorphy, 93
family of curves of genus g, 39
family of stable curves of genus g, 39
family over the Teichmüller curve, 41
flat, 50
flow, 71

Gauß-Manin connection, 60
geometrically imprimitive, 12
geometrically primitive, 12
group of affine biholomorphisms, 9
group of relative periods, 6
group of translations, 9

Hodge norm, 37
Hodge structure, 51

integral lattice, 51

Kontsevich-Zorich cocycle, 74, 76

local system, 43
Lyapunov exponents, 72
Lyapunov spectrum, 72, 75, 76

mapping class group, 5
marking, 3
measurable function, 71
measure space, 71
measure-preserving transformation, 71
meromorphic, 65
modular, 23
moduli space of compact Riemann surfaces, 5
monodromy representation, 45

normal, 18

origami, 11
origami curve, 12

period datum, 121

period domain, 55
period mapping, 61
period matrix, 56
polarization, 52, 59
primitive, 13
probability space, 71
pure, 53

real
    Hodge structure, 51
real variation of Hodge structures, 57

saddle connections, 6
stable curve, 5
Strebel, 19
symplectic map, 32

Teichmüller curve, 11
Teichmüller disk, 10
Teichmüller embedding, 10
Teichmüller geodesic flow, 8
Teichmüller space, 3
translation covering, 6
translation surface, 6
trivial Lyapunov exponent, 75
trivial subrepresentation, 92

unipotent monodromy, 64

variation of Hodge structures, 58
Veech cover, 9
Veech group, 9
Veech surface, 9

width, 98